Soils and Soil Management

Soils and Soil Management

Charles D. Sopher
Jack V. Baird

Reston Publishing Company, Inc.
A Prentice-Hall Company
Reston, Virginia

Library of Congress Cataloging in Publication Data

Sopher, Charles D
Soils and soil management.

Includes bibliographies and index.
1. Soil science. I. Baird, Jack Vernon.
joint author. II. Title.
S591.S783 631.4 77-17335
ISBN 0-87909-801-5

A Prentice-Hall Company
Reston, Virginia 22090

10 9 8 7 6 5 4

Printed in the United States of America

To
Sharron W. Sopher and
Leota P. Baird
for their assistance and understanding

Contents

Preface

This book was developed to be used as a very general soils text for students not having the geological, chemical, and mathematical backgrounds necessary to use the standard four-year college textbooks presently available. It evolved from a set of notes for the basic soils course in the Agricultural Institute program (two-year curriculum) at North Carolina State University. Information deemed necessary for a basic understanding of soils is presented in each chapter. Appendixes follow those chapters where additional information is helpful but not absolutely necessary for the beginning student. Essay questions are presented at the end of each chapter. These questions stress the points needed to understand subsequent chapters. Additional references are cited at the end of each chapter.

It is our intent to try and cover many aspects of soils rather than to present only an agricultural emphasis. Where applicable, examples of soil uses in such areas as waste management, engineering, and urban development are presented. Although we have tried to reduce regional emphasis, our backgrounds have probably contributed an appreciable amount of midwestern and eastern bias. We did not attempt to cover the calcareous and sodic soils of the western United States.

We are grateful to those who have contributed their time in reviewing and editing these materials. It is our hope that this book will serve as a text and reference for the student in technical agriculture.

Instructor's Note

Although laboratory exercises were developed to parallel this text, they were not published because we feel that instructors using the text should develop exercises that fit their particular locality. From several years' experience, we have found that "on site" soils laboratory exercises, when possible, are much more successful than moving the soil into the laboratory. When it is necessary, due to weather, distance, or other circumstances to present "in-house" exercises, we find slides and *fresh* soil samples to be very helpful. The use of soil monoliths early in the course creates many questions regarding "how to get the soil to stick on the board," but generates little true soils interest. However, monoliths are quite valuable once the student understands soil morphology and formation.

The arrangement of this text is as follows: introduction, terminology, soil formations, physical properties, chemical properties, fertilizers, soil classification, and soil conservation. This order may be changed. However, because later chapters often refer to earlier sections, rearrangement could cause the student to check unassigned portions of the text.

1

The Importance of Soils

"The wealth of a nation lies in her soils and their intelligent development."[1]

Soils are one of the most valuable natural resources of a nation. In agricultural production they are an integral part of the ecological system which produces our food and fiber. For agricultural crops, soils serve as a medium which is capable of physically supporting plants as well as acting as a storehouse for the water and nutrients essential for plant growth. For livestock production soils furnish much of the feed necessary to produce the high quality animal products consumed in the world today.

Although it is a common practice to associate the importance of soils with agricultural production and forestry, soils are also very important in urban development, engineering, and recreation. In urban areas soils act as the foundation materials for houses, streets, and buildings. They also work as purification systems for septic tank effluent, and again they are a medium for the growth of lawns, shrubs, and gardens. Many urban areas in the United States have severe problems with cracking building foundations and driveways, overflowing septic tanks, and barren lawns. These

[1] Richard Gordon Moores, *Fields of Rich Toil: The Development of the University of Illinois College Of Agriculture* (Urbana: University of Illinois Press, 1970), pp. 136-37. *Note:* This quotation is a paraphrase of Eugene Davenport's statement that the wealth of Illinois is in her soil and her strength lies in its intelligent development. Eugene Davenport, "Address," Dedication Agricultural Building, University of Illinois.

Figure 1-1. Soils play an integral role in agricultural production.

problems could have been avoided by a thorough understanding of the soils and proper planning prior to building.

Soils are also involved in many engineering procedures. Among other things, they are used as roadbed materials, core materials for dams, and filter beds for waste disposal plants. However, there are certain mixtures of soil materials that render the soil worthless for these uses. Unless one understands how all soils will react when used in a certain fashion, some very expensive mistakes can be made.

With an ever-expanding population and a much reduced work week many persons are now pursuing out-of-doors hobbies that were not possible even a decade ago. Because the land areas of the world are physically limited, the pursuit of these recreational endeavors is often stifled due to lack of space and suitable environment. Unless we fully understand the potential hazards of soil pollution and the soil's ability to cope with pollution caused by the encroachment of man, out-of-doors recreation will soon be gone. In the near future it will be the full responsibility of society to completely protect and conserve soil areas we now treat with indifference.

Figure 1-2. Soils play an important role in on-site sewage disposal. Unless corrected, this septic tank poses a severe health hazard.

In summary, it can be said that soils play an important and integral part of our everyday life. They help supply our food, support our homes and highways, act as building materials for our construction, absorb our waste materials, and give us many beautiful and scenic areas for relaxation and recreation. For these reasons, it is the responsibility of everyone to have an appreciation

Figure 1-3. Soils are a necessary consideration in most engineering projects.

and understanding of our soils, so that all people can conserve and enjoy this valuable natural resource.

REVIEW QUESTION

How do soils affect our everyday lives?

REFERENCES

Brady, Nyle C. *The Nature and Properties of Soils*. 8th ed. New York: Macmillan Co., 1974, pp. 1-8.

2

What Is a Soil?

"What is a soil?"

"Well, there's ground, sod, terra firma, loam, dirt, earth, firmament, organic stuff, mineral matter, stuff plants grow in–anybody knows what a soil is!"

Whenever you attempt to understand something more fully it is quite important that you first define and describe the object you are studying. The goal of this section is to delineate a *soil* and explain it with some meaningful terms.

WHAT IS A SOIL?

Although there are nearly as many definitions of a soil as there are soils textbooks, a widely accepted definition today was devised by the Soil Conservation Service in 1975.[1] This definition basically states that soil is the collection of natural bodies on the earth's surface containing living matter and supporting or capable of supporting plants. Its upper limit is air or water and at its lateral margins it grades to deep water or barren areas of rock or ice. Its lower limit is the most difficult to define, but is normally thought of as the lower limit of the common rooting depth of the native perennial plants, a boundary that is shallow in deserts and tundra and deep in the humid tropics.

---

[1] *Soil Taxonomy: A Basic System of Soil Classification for Making and Interpreting Soil Surveys*, Agriculture Handbook No. 436 (Washington, D.C.: Soil Survey Staff, Soil Conservation Service, U.S. Department of Agriculture, Dec. 1975), p. 1.

This definition tells us that a soil is a collection of natural bodies, or that it is something that is developed by nature. It further tells us that these natural bodies occur on the earth's surface and are capable of supporting living matter and plants.

From this definition we find that the upper and lateral boundaries of this body are fairly clear but that the lower limit is vague and often hard to define. Although this vagueness has been the subject of many heated discussions in academic circles, the student of elementary soils need not be greatly concerned. Soil usage will usually dictate the lower boundary. For example, when evaluating a soil for corn production a corn root zone of up to 60 inches will probably be considered. When evaluating this same soil as a potential home site, the soil should be observed to a greater depth so that the effects on foundations and septic tanks can be evaluated.

SOIL TERMINOLOGY

As a soil or natural body develops on the landscape, distinct layers or bands parallel to the earth's surface may form. These layers or horizontal bands are called soil horizons. Because similar horizons may occur in many different soils, a uniform naming system for soil horizons has been devised. This system consists of calling the uppermost layer where organic materials have accumulated, the O horizon. Beneath the O horizon we find a mineral horizon where clay and nutrients have been removed by leaching and organic stains and residues from the O horizon have accumulated. This horizon is known as the A horizon. It is often called the zone of loss because of clay and nutrient removal.[2] The A horizon will be the upper horizon in tilled fields where the O horizon cannot form or has been removed.

Below the A horizon, in the zone where the materials from the A horizon have accumulated, we find the B horizon or the zone of accumulation. The horizon underlying the B horizon usually contains fresh materials from which the A and B horizons were formed or other materials which have not been affected by soil formation. This is designated the C horizon or the horizon of parent materials. If the C horizon grades to hard rock it becomes an R horizon.

When the need arises, the major O, A, B, and C horizons are

---

[2] The term *zone* is quite general, meaning any level within the soil. Horizons are specifically defined soil layers.

further divided by using the Arabic numerals 1, 2, or 3 after the O, A, B, or C. The O horizon is subdivided into O1 and O2. In the O1 horizon the organic remains are undecomposed and identifiable. The organic remains in the O2 horizon have decomposed to the point that they are no longer identifiable. In the A horizon of older soils it is common to find a dark organic matter rich surface layer. This layer is called the A1 horizon or the Ap horizon if it has been plowed. Under the A1 or Ap horizon of older soils it is common to find a light-colored bleached horizon which has very little clay and few nutrients. This light-colored zone is called the A2 horizon and is labeled the zone of maximum loss. If the A horizon tends to grade toward a B horizon, an A3 horizon may be present. The A3 designation simply means there is an identifiable soil zone which is grading toward a B horizon but is still quite similar to either the A1 or A2.

In the B horizon it is common to find B1, B2, and B3 subhorizons. The B1 is a gradational horizon which is like both the A and B horizons but enough like B to be called a B1 rather than an A3. The B2 horizon represents the middle of the zone of accumulation and is often called the zone of maximum accumulation. The B3 horizon indicates that the lower boundary of B horizon has been reached and that the soil is grading toward the parent materials. When the C horizon is subdivided the numbers 1, 2, 3, etc. are used without special meaning and simply denote C horizon subdivisions.

In highly detailed studies the letters 1, 2, or 3 may be again added to give finely subdivided horizons such as A21, A22, B21, B22, B23, and so forth. It is also common to note horizon features with small letters after the numbers. For example, a B22 g horizon would be a horizon in the middle of the B2 horizon which contains many gray colors associated with wetness. For a complete discussion of soil horizon designations, the student is referred to *Soil Taxonomy*.[3]

It is common for the beginning soils student to assume that all soils must contain O, A, B, C, and R horizons. This is an invalid assumption since many young soils contain only some accumulated organic matter over parent materials. In this case only an A and a C horizon would be present. In highly eroded areas and recent floodplains, the soil may not show any horizonation, i.e., contain any defined horizons.

---

[3] *Soil Taxonomy*, pp. 459-62.

A vertical slice of the earth's crust showing all of the horizons present and their thickness is called a soil profile. Figure 2-1 is a schematic depiction of some of the landscape relationships between the soil profile and associated horizons; Figure 2-2 shows a soil containing several horizons.

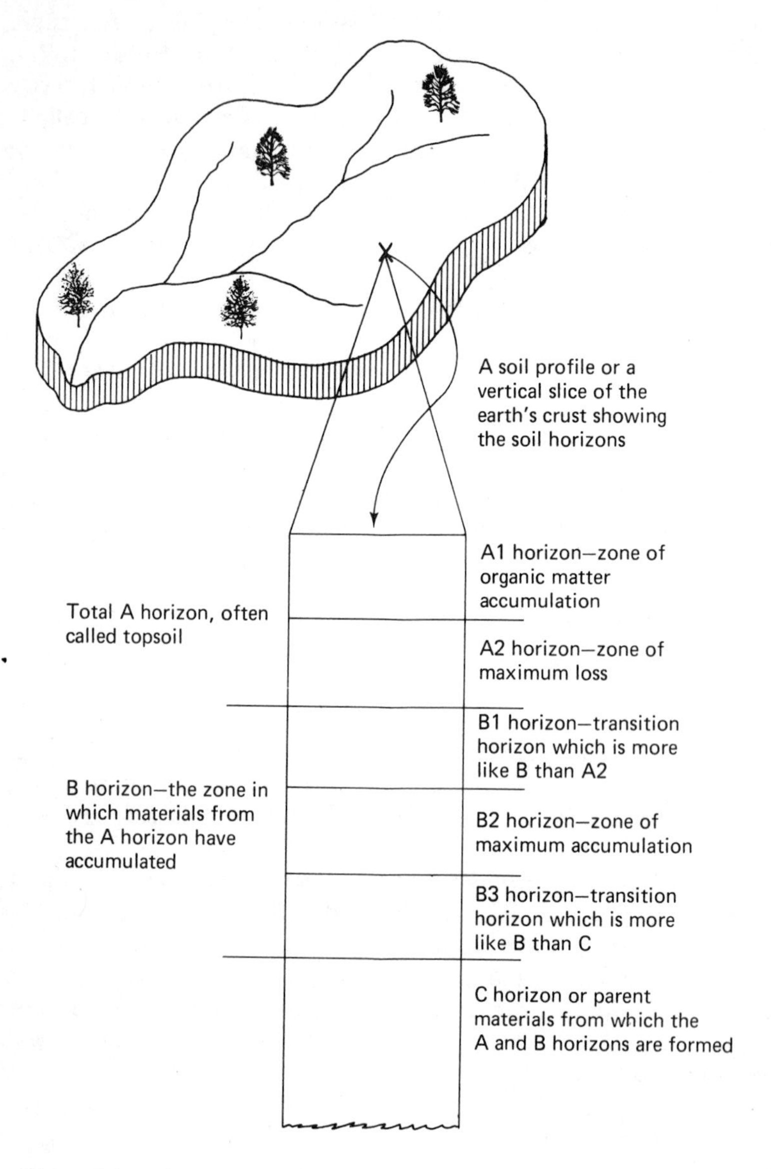

Figure 2-1. A soil profile and its relationship to the landscape.

Figure 2-2. This soil profile is more subtle than the schematic presentation in Figure 2-1. Still, at least A1, A2, B, and C horizons should be recognizable.

THE SOIL BODY (PEDON)

Although the soil profile is a very useful tool in describing a soil it only represents a point on the landscape. It does not give any information on irregularities in horizon thickness, nor does it note intermittent horizons. To overcome the problems associated with studying a soil profile, soil scientists often study a volume of soil known as a *pedon*. The pedon is the smallest volume that represents a soil. Since some soils are more variable than others the lateral dimensions of the pedon are variable enough to account for the variations in horizons. The pedon and its relationship to the landscape are shown schematically in Figure 2-3. Since it is nearly impossible to remove a block of soil as in Figure 2-3, soil scientists will study the sides of a soil pit as shown in Figure 2-4.

Soil scientists often define a soil simply as a collection of similar pedons or as a polypedon. The word *polypedon* simply means more than one pedon.

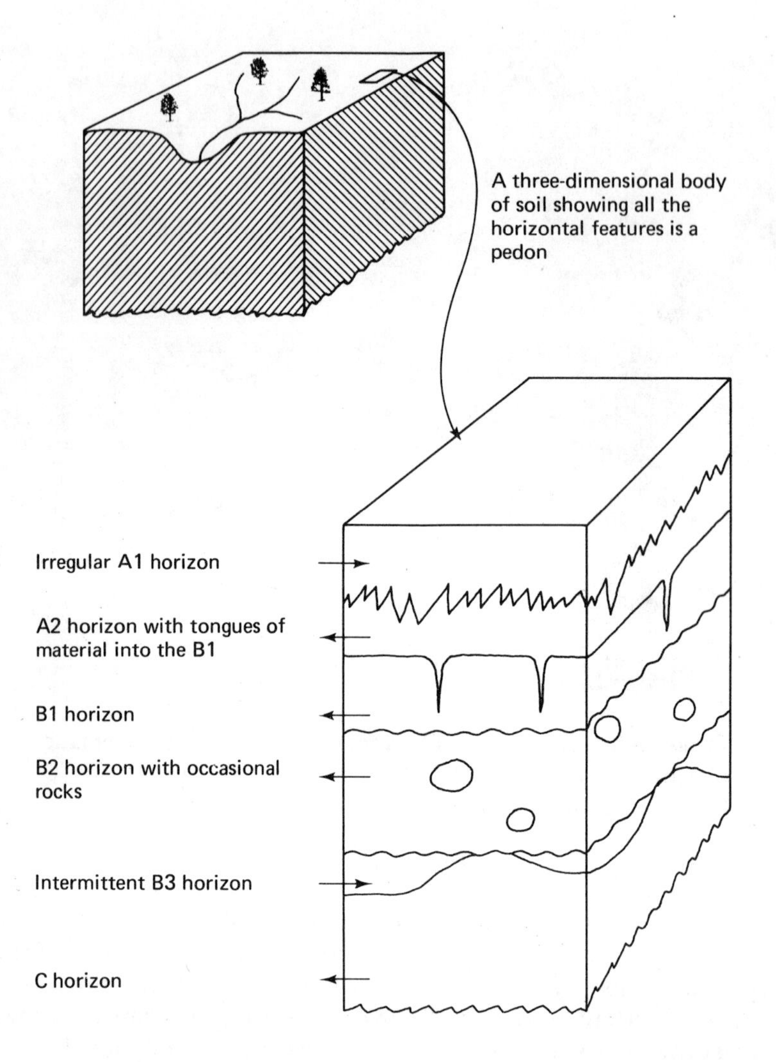

Figure 2-3. The pedon and its relationship to the landscape.

Although soils tend to be variable in horizon thickness and other properties as one traverses the landscape, it is a standard practice to classify soils into groups called soil series. Soil series are simply groups of pedons whose properties fall into a certain range. When the landscape changes so that the pedons no longer fit within the prescribed ranges, they are placed into another grouping or another soil series.

Figure 2-4. A representative pedon may be observed by studying the walls of a pit large enough to encompass the variability of a given soil.

SUMMARY

The materials given in this section represent some of the technical jargon or terminology of the soil scientist. In order to study soils, the student of elementary soils should become quite familiar with these terms and try to derive a mental picture of each item. The new terms given in this section are as follows:

soil
soil horizon
O horizon
A horizon
B horizon
C horizon
parent materials
O1 horizon
O2 horizon
A1 horizon
Ap horizon
A2 horizon
zone of maximum loss
zone of maximum accumulation
B1 horizon
B2 horizon
B3 horizon
soil profile
pedon
polypedon
soil series

REVIEW QUESTIONS

1. What defines a soil's boundaries?
2. Where is the zone of maximum loss located?
3. Parent materials are found in which horizon?
4. Can similar soil horizons occur in different soils? Why or why not?
5. What is a soil profile?
6. What is the total A horizon often called?
7. What is the smallest volume that represents a soil?
8. Soils can be classified into groups called what?
9. Which zone commonly contains a rich layer of organic matter?
10. Do all soils contain O, A, B, and C horizons? Why or why not?

REFERENCES

Berger, Kermit C. *Introductory Soils.* New York: Macmillan Co., 1965, pp. 1-15.

Brady, Nyle C. *The Nature and Properties of Soils.* 8th ed. New York: Macmillan Co., 1974, pp. 7-11, 312-17.

Buol, S. W., Hole, F. D., and McCracken, R. J. *Soil Genesis and Classification.* Ames: Iowa State University Press, 1973, pp. 16-20.

Donahue, Roy L., Shickluna, John C., and Robertson, Lynn S. *Soils: an Introduction to Soils and Plant Growth.* 3rd ed. Englewood Cliffs, N.J.: Prentice-Hall, 1971, pp. 131-35, 149-151.

Foth, H. D., and Turk, L. M. *Fundamentals of Soil Science.* 5th ed. New York: John Wiley & Sons, Inc., 1972, pp. 1-8, 203-07.

Soil Survey Manual. Agriculture Handbook No. 18: Supplement. Wash., D.C.: U.S. Department of Agriculture, May 1962, pp. 173-88.

3

Soil Formation

"Can you define soil formation?"

"Parent material, climate, relief, and vegetation all working overtime—soil formation."

The study of soil formation is called *soil genesis*. Interpreted literally soil genesis means explaining the origin of soils. To the student of elementary soil science the idea of studying soil genesis may seem rather unrelated to modern problems until one remembers that each subject, whether it be history, art, religion, or engineering, must return to the beginnings to provide the foundation for the present. A thorough understanding of soil formation processes is a valuable tool to use in interpreting soils for specific uses.

In Chapter 2 it was stated that a soil is a natural body. This indicates that it was formed by the actions of nature. The agents or factors of nature that are recognized as being responsible for soil formation are:

Parent Materials
Climate
Relief or Topography
Vegetation
Time

An easy way to remember these five factors is to memorize the following sentence: The type of soil developed depends on the amount of time a parent material on a specific topography is exposed to the effects of climate and vegetation.

In order to understand the five factors of soil formation each factor will be discussed separately. Although these factors are discussed separately they do not act independently. The type of soil formed will be the net result of all five factors working at the same time. After discussing each of the factors, an attempt will be made to show how they work together at times and against each other at other times to give rise to this natural body we call a soil.

SOIL PARENT MATERIALS

In Chapter 2, *parent materials* are defined as the materials underlying the soil (in the C horizon) and from which the soil develops in most cases. Discussion later in this chapter will indicate how the materials in the C horizon can be different from those in which the A and B horizons developed.

There are five general categories of parent materials from which a soil may develop:

1. Minerals and rocks
2. Glacial deposits
3. Loess deposits
4. Alluvial and marine deposits
5. Organic deposits

Of these five categories, minerals and rocks are probably the most important on a worldwide basis. They are important because they are the materials that weather or break down to form soils.[1]

Minerals

A *mineral* is defined as a naturally occurring inorganic body that has a fairly definite internal structure and composition which results in fairly definite physical and chemical properties. A *rock* is simply a complex mineral aggregate. Although soils are seldom developed from a deposit of pure minerals, minerals are quite important since they are the component parts of a rock. Geologists have identified several thousand naturally occurring minerals in

[1] In this portion of the text the words *weather* or *weathering* will be used in a geological context to mean the breakdown or disintegration of rocks and minerals at or near the earth's surface by the actions of nature.

the earth's crust. In the following discussion, we will only concern ourselves with those broad groups of minerals which are of prime importance in soil development.[2]

FELDSPARS: Feldspars are the most abundant group of minerals in the earth's crust. These minerals contribute a large amount of potassium and lesser amounts of sodium and calcium to the soil. The most common mineral of the feldspar group that is found in soils is called orthoclase. Two other feldspars, albite and plagioclase, are also found in soils but in much smaller amounts than orthoclase. When feldspars become very weathered they also contribute to the formation of clay.

AMPHIBOLES AND PYROXENES: The amphiboles and pyroxenes are easily weathered groups of minerals that supply calcium, magnesium, sodium, and some iron to the soil. Compared to the feldspars, these minerals are not nearly as common in the earth's crust. Hornblende is a common mineral in the amphibole group and augite is a common pyroxene. Although these minerals are not as abundant in the earth's crust as feldspars, they are large contributors of the above elements.

MICAS: Micas are common minerals in soils. They are characterized by their sheetlike structure and their almost clear appearance when present in thin sheets. The main contribution of mica to the soil is the potassium it releases as it weathers and the sheets break down and come apart. The most common mica is muscovite, or white mica. Black mica, or biotite, is also present in soils but in much smaller amounts than muscovite. In older soils it is not common to find biotite since it weathers away much faster than muscovite.

SILICA: The silica minerals are extremely important since quartz (the most abundant of the group) or sand is the most abundant mineral in many soils. The silicas are extremely hard minerals that resist weathering and are left in the soil after minerals in the previous groups have weathered away completely. Two other silica minerals which are of mild importance in soils are cristobalite and chalcedony (flint).

For many years it was thought silica was unimportant in plant growth. However, recent studies indicate it is beneficial to some plants.

IRON OXIDES: There are four iron oxides found in the soil.

---

[2] Kermit C. Berger, *Introductory Soils* (New York: Macmillan Co., 1965), pp. 23-33.

Hematite and limonite are iron-rich minerals found in fairly large quantities. Two other iron oxides, goethite and magnetite, are also found, but in smaller quantities. These minerals furnish iron for plant growth and are responsible for the red and yellow colors found in many southeastern soils.

CARBONATES: Carbonates are easily weathered minerals which have long been weathered and leached from many soils. In more arid desert regions they may only partially leach and then accumulate in certain soil horizons. In much of the Corn Belt, the parent materials contain calcium carbonates and give rise to non-acid soils.

The two most important carbonate minerals found are calcite and dolomite. Calcite is a calcium limestone while dolomite is a calcium magnesium limestone. These two minerals comprise much of the agricultural limestone used in the world.

OTHER MINERALS: The following minerals are ones which have been extracted from their general group since they are the only ones in the group which are of prime importance in soils.

1. *Gibbsite* is an aluminum-rich mineral that contributes greatly to the large amount of aluminum found in many soils.

2. *Apatite* is a member of the phosphate group. It is not a highly abundant soil mineral but it is an important source of phosphorus.

3. *Tourmaline* is a mineral that is found only in small amounts in the soil. It is quite important as it is the only source of soil boron.

4. *Zircon* is a mineral that is quite hard and weathers very slowly in the soil. It has no value as a plant nutrient but is often used as an index of soil age. Because it is so resistant to weathering, old soils have a larger relative amount of zircon compared to young soils.

5. *Pyrite* is an iron mineral that provides both iron and sulfur in the soil. It is a shiny yellow mineral and is often called fool's gold.

6. *Gypsum* is an easily weatherable mineral that is weathered out of soils similar to calcite. It is a calcium

sulfate material which supplies both calcium and sulfur to the soil. In the peanut-producing areas it is used as land plaster (a source of calcium).

The minerals previously discussed are often called primary minerals. These are minerals formed in nature as a result of certain elements coming together and crystallizing. As these primary minerals (especially feldspars and micas) weather, certain clay minerals form. These clay minerals are called secondary minerals and are quite abundant in nearly all soils. The common clay minerals found in soils are kaolinite, montmorillonite, vermiculite, and illite. Because these clay minerals are responsible for many soil properties, they will be discussed in detail later.

In summary, the minerals important in soils and soil formation are presented in Table 3-1. This table also indicates the contribution to the soil of the mineral or group of minerals.

Rocks

Earlier in this chapter it was indicated that most soils are actually formed from rocks rather than pure mineral deposits. Minerals were discussed first since they are the major components of rocks and should be understood before a discussion of rocks can take place.

Rocks are usually classified by placing them into one of three groups, depending on their mode of formation. These three groups are igneous, sedimentary, and metamorphic rocks.

IGNEOUS ROCKS: Igneous rocks are rocks formed from the cooling of molten materials which have been pushed up from the center of the earth. These molten materials fall into one of the following two categories: lava or magma. Lava is material which comes to the surface of the earth and is cooled quickly. Magma, the more dominant of the two rock materials, is formed when molten materials are pushed part way to the surface of the earth and then cool very slowly. It is this slow cooling of the magma that allows for many of the crystalline mineral structures observed in nature.

Although geologists have recognized many different kinds of igneous rocks, only seven will be discussed. These seven igneous rocks along with their average mineral composition are presented in Table 3-2.

Table 3-1
Some Important Minerals
and Their Contribution to the Soil

Mineral Group and Species	*Contribution to the Soil*
Feldspars	
(1) Orthoclase	Potassium, forms clay
(2) Albite	Sodium, forms clay
(3) Plagioclase	Sodium, calcium, forms clay
Amphiboles	
(1) Hornblende	Calcium, magnesium, sodium, some iron
Pyroxenes	
(1) Augite	Calcium, magnesium, some iron
Micas	
(1) Muscovite	Potassium, forms clay
(2) Biotite	Potassium, forms clay
Silica	
(1) Quartz (2) Cristobalite (3) Chalcedony	All are resistant to weathering and are the main component of sand
Iron Oxides	
(1) Hematite (2) Limonite (3) Goethite (4) Magnetite	All contribute iron to the soil and are responsible for the red and yellow colors in the B horizons of well drained soils
Carbonates	
(1) Calcite	Calcium } Common limestones
(2) Dolomite	Calcium, magnesium } Common limestones
Other Minerals	
(1) Gibbsite	Aluminum
(2) Apatite	Phosphorus
(3) Tourmaline	Boron
(4) Zircon	Resistant to weathering, used to index soil age
(5) Pyrite	Iron and sulfur
(6) Gypsum	Calcium, sulfur
Clay Minerals	
(1) Kaolinite (2) Montmorillonite (3) Vermiculite (4) Illite	Found accumulated in most soils. Give rise to most of the chemical and physical properties observed in soils

Table 3-2
Selected Igneous Rocks
and Their Mineral Components

Rock	*Mineral Components*[1]
Granite and rhyolite	50% feldspars 30% quartz 20% iron and magnesium minerals[2]
Basalt and gabbro	50% feldspars 50% iron and magnesium minerals
Diorite and andesite	75% feldspars 25% iron and magnesium minerals
Obsidian	60% feldspars 30% quartz 10% iron and magnesium minerals

[1] The mineral compositions given are only averages. For example, in granite the feldspar content may range from 35% to 65%.

[2] These minerals containing both iron and magnesium are commonly called ferromagnesian minerals. This is a general term for dark-dolored minerals such as biotite, augite, ad hornblende.

Within the igneous rock group, the rocks are further classified according to their color and texture. The color of an igneous rock is dependent on the minerals from which it is formed. In general, dark-colored igneous rocks contain dark-colored ferromagnesian minerals such as biotite, augite, and hornblende. Gabbro and basalt are good examples of dark-colored igneous rocks. Light-colored igneous rocks contain a high amount of quartz. Examples of these light-colored igneous rocks are granite and obsidian.[3]

The texture of igneous rock is determined by how fast the molten materials cool and how large the mineral crystals grow within the rock. Obsidian and basalt are examples of fine-textured igneous rocks. Granite and gabbro are coarse in texture.

The color and texture of igneous rocks are very important

[3] It has been quite common in the past for geologists and soil scientists to refer to light-colored igneous rocks as "acid" rocks and dark-colored igneous rocks as "basic" rocks. These terms were coined because the soils formed from the lighter-colored high quartz igneous rocks were more acid and lower in nutrients than those derived from darker-colored rocks. Today the terms *acid* and *basic* rocks have been replaced by the terms *felsic* and *mafic* rocks, respectively. This was done to avoid confusion since many people interpret an acid rock to mean a rock that was acid rather than one that formed an acid soil. Also basic rocks seldom form truly basic soils.

factors in soil formation. Since the dark-colored igneous rocks are high in minerals which contain nutrients such as calcium, iron, magnesium, and potassium, the soils formed from these rocks also contain these nutrients. Likewise, soils formed from light-colored igneous rocks will be high in quartz and low in nutrients. Similar conclusions can be drawn from the rock textures. Fine-grained rocks will give rise to soils containing fine materials while coarse-textured ones will develop into soils which are very coarse. In Table 3-3 some of the important soil-forming rocks are grouped by their texture and structure.

Table 3-3
Important Soil-Forming Igneous Rocks

	Color		
Texture	*Dark* ⟶	*Intermediate* ⟶	*Light*
Coarse ↓	Gabbro	Diorite	Granite
Intermediate ↓	Basalt	Andesite	Rhyolite
Fine		Obsidian	

Using the information in Table 3-3, the student can conclude that soils derived from granite will be coarse in texture, high in quartz, and low in nutrients, whereas soils derived from basalt will be higher in nutrients, low in quartz, and of a medium texture. The beginning student should not overlook the broad soil interpretation that can be made by understanding the composition of igneous rocks and the type of soil materials derived from them.

SEDIMENTARY ROCKS: Sedimentary rocks are formed by the bringing together of sediments and small rock fragments and by their cementing, either chemically or by compression. Although these rocks comprise only 5 percent of the rock volume of the outer ten miles of the earth's crust, they cover 75 percent of the surface.

The process of bringing together sediments is called sedimentation. The agents responsible for this process are wind, running water, and precipitation.[4] The methods by which these agents work are best explained by examples. If a strong wind passes over a barren area it may pick up sand and soil particles and

[4] Precipitation takes place when minerals are dissolved by water at one location and later crystallized out at another location.

carry them away. The wind will carry these particles as long as it maintains its velocity. However, as climatic patterns change, the wind may slow down. As it slows it will first drop the largest particles it is carrying. Finally it will stop and even the fine dust particles will settle out. If this wind pattern prevails for a long period of time the deposits from it may become quite thick. As the deposits become thicker they may be cemented into sedimentary rocks.

Sedimentary deposits may also be formed by running water's transporting materials much the same way as wind. Water is capable of moving much larger particles than wind. During severe floods, water running at high velocities can move small rocks and coarse gravel great distances.

The process of precipitation takes place when water moves downward through the earth's surface and dissolves minerals such as calcite, dolomite, and gypsum. As long as fresh water is moving through the surface, these minerals will keep dissolving. If, however, the water stops moving the dissolved materials may become so concentrated that they recrystallize or precipitate out and form a sedimentary deposit.

Whenever the sedimentary products of wind, water, or precipitation are cemented together, a sedimentary rock is formed. The composition of the rock formed will depend on the type of material which was transported. Usually the sedimentation process is not continuous and takes place only during wind storms, floods, and soaking rains. This causes the sediments to be deposited in distinct layers. When these sediments are then cemented, the resulting sedimentary rocks show a very distinct layering effect.

The sedimentary rocks most commonly involved in soil formation are: limestones, sandstones, shale, and conglomerate. These rocks along with their mineral composition and characteristics are given in Table 3-4. Because 75 percent of the earth's surface is covered by these rocks, this table should be studied in detail.

METAMORPHIC ROCKS: Metamorphic rocks are igneous or sedimentary rocks which have been subjected to enough heat, chemical activity, and/or pressure to radically alter their characteristics. This process, known as metamorphism, is usually associated with volcanic activity during magma formation or during times when the earth's crust is changing shape and mountains are being formed. It is beyond the scope of this text to go into the processes of metamorphism. However, the metamorphic rocks important in soil formation are presented in Table 3-5. The soils formed from these metamorphic rocks are very similar to those formed from the original igneous and sedimentary parent rocks.

Table 3-4
Chatacteristics of Some Common Sedimentary Rocks

Rock	*Minerals Present*	*Rock Characteristics*	*Soils Formed*
Limestone	Dolomite or Calcite	Variable in color from white to gray to light red and yellow	Basic soils high in calcium and magnesium
Sandstone	Mainly quartz	Variable, white to red to brown	Acid soils which will leach quite easily and be low in nutrients
Shale	Clay minerals, some quartz and organic matter	Layered structure, variable color depending on minerals present	Variable depending on amounts of quartz and ferromagnesian minerals present
Conglomerate	Variable	Cemented gravel and rock fragments larger than 2 to 4 mm	Variable but usually coarse and contain much gravel

Table 3-5
Common Soil-Forming Metamorphic Rocks

Metamorphic Rock	*Original Rock*
Quartzite	Sandstone
Slate	Shale
Marble	Limestone
Schist	Basalt
Gneiss	Granite

ROCK-WEATHERING PROCESSES: Weathering was defined earlier in this chapter as the breakdown or disintegration of rock at or near the earth's surface, by the actions of nature. Since we now have defined some of the hard rocks that may be present at the earth's surface, we will take a closer look at the processes which actually break them down.

The forces of weathering may be either mechanical or chemical processes. Mechanical forces consist of some force in nature physically breaking the rock into smaller pieces. Examples of these forces are temperature, wind, ice, water, and plant roots. Rapid temperature changes can cause expansion and cracking of rocks. Wind may literally sandblast the rock surfaces away. Ice causes rocks to break up by expanding in crevices and cracks. Ice in large amounts (glaciers) can grind rocks into dust if it is forced to slide over them. Running water is quite abrasive as it either moves sediments over rocks or as it carries rocks themselves downstream. As plant roots grow they may extend into cracks in rocks and cause breaking and further cracking. Although these mechanical forces operate quite slowly, they are very effective over long periods of time.

The specific processes which chemically disintegrate rock are beyond the scope of this text.[5] Instead, the following general discussion is presented. Water is the solution responsible for much chemical weathering. It is capable of reacting with rock surfaces and causing them to change chemically. Water also simply dissolves many minerals. Furthermore, it can enter into oxidation reactions and cause disintegration. In addition to water, there are numerous weak organic and inorganic acids which can cause rock breakdown.

The beginning student should remember that weathering is as endless as time itself. It is a process in nature that never stops. By understanding this process it is easy to see how many of the earth's features and its soils were formed.

THE ROCK CYCLE: In order to better understand the processes of rock formation and weathering, the rock cycle is shown in Figure 3-1.

This cycle tells us only what *can* happen. It does not say this *will* happen. For example, igneous rocks may be changed to meta-

---

[5] Nyle C. Brady, *The Nature and Properties of Soils*, 8th ed. (New York: Macmillan Co., 1974), pp. 279-287.

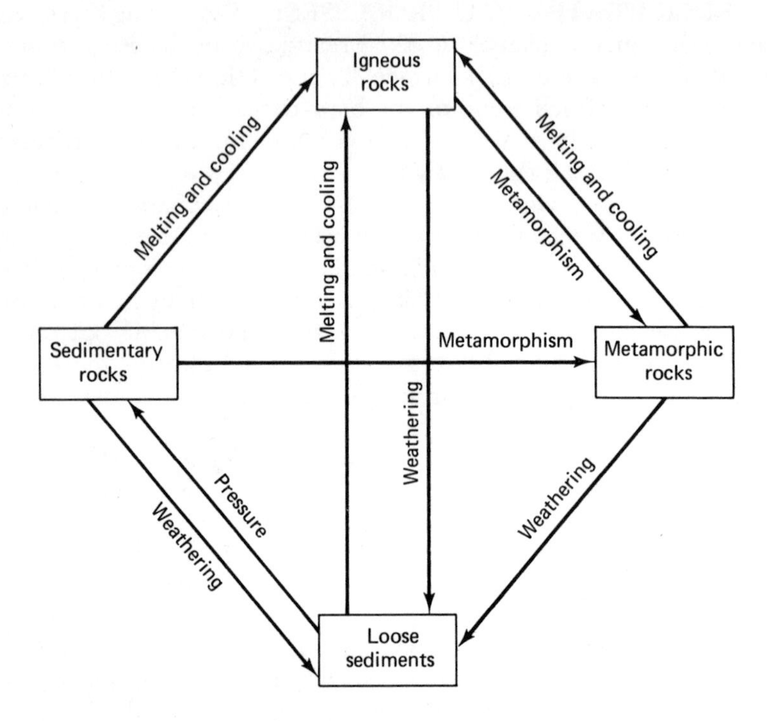

Figure 3-1. The rock cycle.

morphic rocks, they may weather to sediments, they could be remelted, or they could simply remain under the earth's surface as unweathered igneous rock. One thing all the rocks in this cycle do have in common is that they can weather to loose sediments. It is in these loose sediments that soils form.

Alluvial and Marine Deposits

Alluvial deposits usually refer to sediments carried by and deposited in fresh water. Marine deposits are sediments deposited in the ocean. To the beginning student this distinction between deposition in fresh as compared to salt water may seem quite trivial. It is important, however, because the parent materials and resulting soils are usually quite different.

In this text we will discuss three types of alluvial deposits which are common soil parent materials. These three types of alluvial deposits are alluvial fans, floodplains, and deltas.

As a stream moves down a mountain it attains a high water velocity and is capable of moving large amounts of sediments. When this stream reaches a flat valley it suddenly loses its velocity

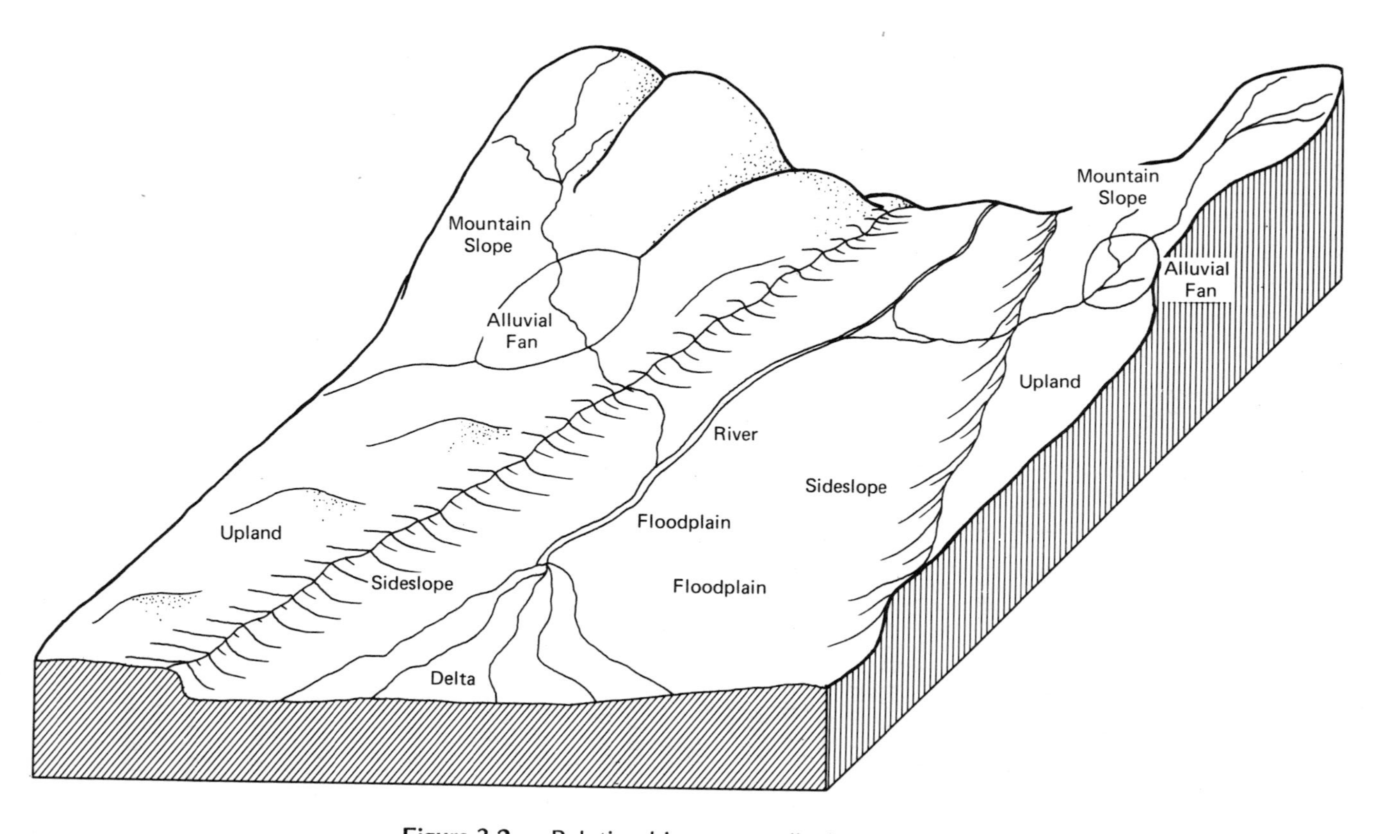

Figure 3-2. Relationships among alluvial deposits.

and, consequently, many of the sediments it is carrying settle out at the foot of the slope. The areas where these sediments are dropped are called alluvial fans. These alluvial fans are quite common at the foot of mountain slopes in both the Appalachian and Rocky Mountains.

Generally the soils on alluvial fans are well drained. The composition of these alluvial fan soils is dependent on the type of rocks and minerals found on the mountain slopes above the alluvial area.

Although alluvial fans are prominent in mountain areas, probably the most common alluvial deposit is the floodplain. As a stream or river flows down a gentle slope it tends to wander across the landscape in a series of curves. Geologists call this wandering and curving *meandering*. Over a long period of time this meandering will cause the stream to have broad, flat bottomlands beside it. These bottomlands are called *floodplains*. Each time the stream floods and overflows its banks the floodwater will spread over these floodplains. As the floodwaters leave the rushing stream channel and spread over the floodplain they immediately lose their velocity and drop the sediments they are carrying. With successive floodings the old stream channels caused by meandering are filled with sediments and a broad flat floodplain containing sedimentary parent materials is formed.

Floodplains are usually poorly drained and tend to be swampy in many places. The composition of the floodplain sediments will be dependent on the materials from which they were eroded. Because these sediments are often eroded from the A1 horizons of surrounding upland soils, they may contain fairly high amounts of organic matter. They also may be high in nutrients because of erosion from highly fertilized surrounding croplands.

Floodplains are a common landscape feature in many parts of the world. In the United States they contribute vast areas of agricultural soils along the Mississippi, Illinois, Missouri, and Ohio Rivers.

Whenever a river empties into a large body of water and the wave action is not sufficient to keep the river sediments suspended, a delta forms. Deltas are usually swampy and dissected by many small streams. Usually they are subject to frequent flooding and elaborate flood control is needed to farm them. Because most of the coarse material carried by the river is usually deposited upstream on the floodplain, deltas often contain large amounts of clay and fine sediments. The deltas of the Mississippi and Nile Rivers are good examples of this type of alluvial deposit.

Most alluvial deposits are not tremendously old; thus the soils developed on them are still fairly young and unleached. They seldom show strong horizonation and usually do not contain A2 horizons.

If stream sediments are not deposited in the alluvial deposits previously discussed, they will eventually find their way into the oceans. Once in the ocean they will be sorted and the coarse particles will be deposited close to the shore and the clays farther out. Because of wave action and storms these marine sediments seldom become thick enough to rise above sea level. However, as mountains are pushed up and/or ocean levels fall, these sediments are often exposed. Along the Atlantic coast and the Gulf of Mexico, a strip of these marine sediments form a coastal plain as much as 50 to 150 miles wide in many places.

Compared to the alluvial deposits previously discussed, marine sediments are quite old and their soils are well developed and highly leached. Thick A2 and B horizons are common and quartz is often the dominant mineral present. Because of the old age of the coastal plains they are often dissected by fresh water streams and alluvial deposits are present much the same as in any upland area. This stream dissecting results in a landscape with an intricate covering of both young alluvial deposits and old marine sediments which leads to a very complex soil pattern.

Glacial Deposits

A glacier is a large ice mass which flows under the pressure of gravity. It is caused by annual additions of snowfall being greater than the amount of snow melted each year. As the snow accumulates, it is compressed into ice which tends to flow away from the areas of accumulation.

The theories of how glaciers form and the technical names of various glacial features are beyond the scope of this text. We will only be concerned with the extent to which glaciation occurred in the past and the soil that parent materials left behind.

In the last million years there have been four major periods of glaciation in the world. These periods are presented in Table 3-6.

Table 3-6
Periods of Glaciation

Glacial Period	*Age (Years)*
Wisconsin	10,000- 100,000
Illinoian	250,000- 350,000
Kansan	600,000- 750,000
Nebraskan	900,000-1,000,000

These glaciers were massive ice sheets as much as two to three miles thick in many places. At the peak of these ice invasions, the northern one-half of North America, northern Asia, and northern and central Europe were covered. In the United States these glaciers roughly advanced southward to New Jersey, northern Pennsylvania, southern Ohio, southern Indiana, southern Illinois, northern Missouri, eastern Nebraska, eastern South Dakota, and the northern sections of Colorado, Idaho, and Oregon. As these glaciers moved southward they rounded hills and gouged valleys and swept the existing soils and rocks ahead of them. When the glaciers retreated during the interglacial periods these minerals were left behind. These transported rocks and soils then became the parent materials for today's soils.

The material left by the retreating glaciers is generally termed *glacial drift*. If the material was washed away from the receding glacial front by the water from the melting ice, it is called *glacial outwash*. If the glacier simply receded and deposited materials, the remaining deposit is called *glacial till*.

Glacial outwash is often characterized by being coarse and containing much gravel. It also may be stratified (layered) due to the action of water. Glacial tills are often characterized by being much finer and containing much more clay. Because till has not been moved by water, it also may contain a wide range of various sizes of rock.

The soils developed on these glacial drift areas are usually very high in nutrients because original drift contained many mafic rocks. Today these glacial soils comprise a large portion of the midwestern Corn Belt.

Loess Deposits

Loess deposits are deposits of windblown silts. In the United States these deposits occur along the Mississippi River and its major tributaries. Although there is not complete agreement among geologists and some soil scientists about the origin of loess, most agree that it was wind-deposited. The most accepted theory is that during glaciation the large rivers, such as the Mississippi, were quite wide due to meltwaters during periods of glacial recession. When the meltwaters subsided, broad barren floodplains were exposed to the prevailing westerly winds. These winds then swept the silt and finer particles from the floodplains and deposited them along the eastern portion of the Mississippi Valley. The states which received most of this loess were Kansas, Nebraska, Iowa, Illinois, Indiana, western Kentucky, Tennessee, and Mississippi.

It is not within the scope of this text to discuss all the peculiarities of loess and loess-derived soils. However, the student should be aware that these soils exist and that they are productive agricultural soils. He should also be aware of the fact that loess-derived soils have many peculiar engineering properties. They will shift when placed under a stress, and they will slide and flow quite easily when wet. It is wise to obtain technical advice when using these soils for urban and engineering uses.

Organic Deposits

In our previous discussions of parent materials we have been dealing with inorganic minerals deposited by certain actions of nature. There are also naturally occurring organic deposits which act as soil parent materials. Although these organic deposits are not as extensive as the other deposits discussed, they comprise millions of acres scattered over the earth's crust. They are also quite important agriculturally in many areas.

Organic deposits form in swampy and marshy areas. If the water is shallow, sedges, reeds, grasses, and even cypress may grow in it. As these plants die or shed their leaves the remains are submerged in water where they cannot be oxidized. Over many years these unoxidized materials will accumulate and finally fill the area with organic matter. Many shallow potholes left by glaciers were filled with organic deposits in this fashion.

In areas of deep water where plants cannot grow, the organic matter is accumulated from the edges of the lake inward. The vegetation around the lake edges finally forms deposits deep enough to support more plant life, which in turn causes more residue accumulation and eventually the lake is filled.

Organic deposits fall into two general categories: peat and muck.[6] If the organic matter is decomposed to the extent that the original source of vegetation cannot be identified, it is called *muck*. When identifiable remains of the organic residues are present the organic deposit is called *peat*. Peat deposits are further divided into fibrous peats and woody peats. Fibrous peats are those which are formed from fibrous plants such as sedges, reeds, mosses, and grasses. Woody peats are those formed from trees such as pines, sweet gum, and cypress.

Generally an organic deposit is sufficient to form organic

---

[6] *Soil Taxonomy: A Basic System of Soil Classification for Making and Interpreting Soil Surveys*, Agriculture Handbook No. 436 (Washington, D.C.: Soil Survey Staff, Soil Conservation Service, U.S. Department of Agriculture, Dec. 1975), pp. 211-26.

soils or *Histosols* when it is greater than 16 inches in thickness and over 30 percent organic matter.[7] These are only the lower limits for defining organic soils. In many areas these soils are several feet thick and contain nearly 100 percent organic matter.

The major areas of organic deposition in the United States are along the Atlantic coast in North Carolina and Florida, along the Gulf Coast in Louisiana, and in the glaciated areas of the North. In these areas it is not uncommon to find mucks and both types of peat.

In the last two decades organic soils have started being used quite intensely. Many of the fibrous peat areas of the glaciated North are being dug and sold for uses in nurseries, greenhouses, and lawns. These areas are also being used for vegetable crop production to a great extent. In the South and Southeast many large organic deposits have been cleared and drained for farm land. This clearing and draining has caused the organic materials to decompose and the soil levels are actually subsiding. Also many woody peats and certain mucks were cleared and drained only to find that they were not manageable under present technology. These areas may eventually need to be returned to their natural state.

The use of organic soils is relatively new and the technology sparse. If we are to develop these areas intelligently, we must be cautious and not ruin them before we start.

In this text the differences in use and management of organic soils as compared to mineral soils will be indicated whenever necessary. The discussions will be brief because the chemistry involved in organic soils is beyond the scope of this text.

This concludes our discussion of the formation and types of parent materials important in soil formation. We are now ready to study how climate, relief, vegetation, and time affect the types of soils formed from these parent materials.

CLIMATE

In soil formation, climate is a general term for temperature and rainfall. The main contribution of both of these factors is their effects on weathering, organic matter production, and organic matter decomposition.

Temperature plays an important role in chemical and physical weathering of parent materials. Temperature fluctuations can cause

[7] *Histosol* is the technical group name for all organic soils including mucks, fibrous peats, and woody peats.

enough expansion and contraction to crack hard rock. Increased temperature can speed up chemical weathering to a great degree. Temperature also has a direct effect on the amounts of organic matter produced. Within the temperature ranges observed in the world, organic matter production will increase with increasing temperature, provided rainfall is sufficient. Increased temperatures also increase organic matter decomposition. Thus, organic matter accumulation is related to the relative rates of both production and decomposition.

Rainfall affects soil formation by speeding up physical and chemical weathering and controlling organic matter production and decomposition. As rainfall increases, erosion of rocks increases and sedimentation takes place. Increased rainfall also greatly enhances chemical weathering by increasing leaching and the chemical breakdown of certain minerals. When rainfall is high, organic matter production is increased provided temperatures are in the proper range for plant growth. If rainfall is high enough to waterlog the soil, organic matter decomposition will be decreased because of reduced oxidation.

At this point in the discussion of climate, it is probably obvious that the effects of rainfall and temperature are closely related and that both factors must be considered in soil formation. For example, 200 cm (79 inches) of rainfall in a warm climate may cause great amounts of organic matter to be produced, but the warm temperature will decompose it and give little net organic matter accumulation. However, 100 cm (39 inches) of rainfall in a cool climate may lead to great amounts of organic matter accumulation because organic matter decomposition is much slower.

RELIEF OR TOPOGRAPHY

Topography contributes to soil formation by affecting soil drainage and erosion. Soils which form on steep slopes tend to have shallow profiles because erosion takes place nearly as fast as soil formation. On slopes of 25 to 40 percent it is not uncommon to have soils with only A and C horizons. This is because erosion is taking place at a rate fast enough to allow only a shallow A1 horizon to form.

The effect of topography on soil drainage is very important. On slopes and uplands, drainage is usually sufficient to produce well aerated, oxidized soils with bright-colored B horizons and low organic matter contents. In lowlands and depressional areas, poor soil drainage will cause soils to have dark gray B horizons and high organic matter contents in the A horizon. These gray colors and the organic matter accumulation are simply due to less oxidation caused by waterlogging.

VEGETATION

For a simplified discussion of the effects of vegetation on soil formation, we will break vegetation into two categories: forest and grasses. Soils developed under grassland vegetation tend to have thick, dark, organic matter-rich A horizons with organic stains and coatings into the upper B horizons. These dark colors and organic matter accumulations are due to the fact that 50 percent of the organic matter produced by grasses is in the root system. Furthermore, this root system is composed of many fine fibrous roots which are distributed throughout the A and upper B horizons. When these grasses die, this root system is decayed and organic matter and dark organic stains are left behind. Grasses also recycle plant nutrients. That is, they take nutrients from the soil and accumulate them in the tops. Then, when the grass dies, these nutrients are leached back into the soil and are available to be used again.

In contrast, soils developed under forest vegetation (trees) usually have thick O horizons, thin organic A1 horizons, and thick bleached A2 horisons. There is seldom any organic matter staining in the B horizon. This difference is caused by trees having a root system consisting of large roots which go much deeper into this soil. Also the only organic matter deposits trees make to the soil are contributed to the O horizon in the form of leaves and dead limbs. Although these leaves and limbs are added in large quantities, they are not distributed throughout the soil profile and they are not of the succulent nature of grass roots. Neither do trees recycle nutrients as quickly as grasses. Trees do, however, recycle nutrients from a greater depth.

TIME

Time is important in soil formation because it determines the degree to which the other factors express themselves. For example, a soil forming from granite would need a very long time period to develop thick A2 and B horizons, simply because it would be a slow process to break down the granite and form these horizons.

In general we can make the following four statements about the effects of time on soil formation:

1. Older soils have deeper soil profiles.

2. Older soils are usually more highly weathered.

3. Older soils contain thick A2 and B horizons.

4. Older soils have usually lost their plant nutrients due to leaching.

This concludes our discussion of the factors of soil formation. Again, it must be stressed that these factors do not act independently to form a soil. The soil formed is the product of all five of these factors working together and/or against each other. For example, a granite in a humid region, on a flat slope, under hardwood vegetation, may develop into a deep acid soil with distinct A2 and B horizons. Likewise, the same granite on a steep slope may develop only a shallow profile completely lacking an A2 horizon.

With the discussion contained in this chapter in mind, the student should now be able to answer the following question often asked by farmers and land managers: "Why do I have so many different soil series on my land?" The answer is quite simple. If he had three parent materials on five topographies developed under two vegetations, one climate, and two periods of time, he would have the following number of soil series: 3 parent materials × 5 topographies × 2 vegetations × 1 climate × 2 time intervals = 60 soil series.

SUMMARY

Soil formation takes place whenever hard rock weathers and accumulates in sufficient quantities to form a medium for plant growth. This accumulation may take place by the rock weathering, or it can be caused by the actions of wind (loess accumulation), water (alluvial and marine deposits), or ice (glacial deposits). Once these weathering products accumulate, soil formation starts.

Soil formation is initiated by adding organic matter and forming clay in these weathered materials. As leaching occurs clay is moved from the A to the B horizon and an A2 horizon forms. With further weathering most of the soluble minerals are dissolved and leached away. As the soil gets older a thick bleached A2 horizon forms and the iron and aluminum in the B horizon may become segregated from the clay. This segregation can continue and much of the clay may even weather away. In very old age we find an oxidized, leached soil high in quartz and zircon and low in plant nutrients.

Figure 3-3. Thick A2 and B horizons are characteristic of well-developed soils.

REVIEW QUESTIONS

1. Can you define the following new terms?

parent material	hornblende	calcite
topography	augite	dolomite
time	mica	gibbsite

climate	muscovite	apatite
vegetation	biotite	tourmaline
weather	silica	zircon
mineral	quartz	pyrite
rock	cristobalite	gypsum
feldspar	chalcedony	primary minerals
orthoclase	hematite	secondary minerals
albite	limonite	kaolinite
plagioclase	goethite	montmorillonite
amphibole	magnetite	vermiculite
pyroxenes	carbonates	illite
igneous rock	metamorphism	glacial outwash
lava	alluvial fans	glacial till
magma	floodplain	loess
sedimentation	delta	muck
precipitation	meandering	fibrous peats
metamorphic	glacial drift	woody peats
peat		

2. How does each of the five soil-forming factors affect soil formation?

3. What does each of the minerals discussed contribute to the soil?

4. What is the rock cycle?

5. Summarize how a soil forms.

REFERENCES

Berger, Kermit C. *Introductory Soils.* New York: Macmillan Co., 1965, pp. 23-33.

Brady, Nyle C. *The Nature and Properties of Soils.* 8th ed. New York: Macmillan Co., 1974, pp. 277-312.

Buol, S. W., Hole, F. D., and McCracken, R. J. *Soil Genesis and Classification.* Ames: Iowa State University Press, 1973, pp. 76-88, 108-70.

Donahue, Roy L., Shickluna, John C., and Robertson, Lynn S. *Soils: An Introduction to Soils and Plant Growth.* 3rd ed. Englewood Cliffs, N.J.: Prentice-Hall, 1971, pp. 68-100.

Foth, H. D., and Turk, L. M. *Fundamentals of Soil Science.* 5th ed. New York: John Wiley, 1972, pp. 203-30.

4

Soil Physical Properties

"What are soil's physical properties?"

"Soil physical properties are those properties you can see, feel, taste, and smell."

"Soil physical properties are properties that don't require a chemical reaction."

"Soil texture, structure, consistency, color, and permeability."

Modern science has given us many elaborate methods and instruments for measuring soil properties. Although these methods are precise and highly accurate, they are often unavailable to the soil manager making field decisions. When faced with these decisions the manager must rely on what he can tell about the soil by the way it looks and feels.

With a small amount of training, a vast amount of information can be acquired by simply observing soil properties that can be seen, felt, smelled, and tasted. By observing soil color, we can estimate organic matter contents, iron and manganese contents, soil drainage, and soil aeration. By feeling the soil we can estimate the kinds and amount of different sizes of particles present. In various areas it is common to test for salt concentrations by tasting a small amount of suspected salty horizons. Certain humid region soils high in sulfur can be detected by their odor. In addition to these bits of information, it is easy to determine rooting patterns, water movement, and soil aeration by observing the overall soil pedon.

It is the objective of this chapter to discuss those soil properties which contribute to the appearance and feel of a soil. The properties that add to the appearance and feel of a soil are called *soil physical properties*. These are distinguished from chemical properties, since they do not require a chemical reaction to be effective. The major soil physical properties are: soil texture, soil structure, soil consistency, soil color, soil permeability, and overall soil physical characteristics.

SOIL TEXTURE

In previous chapters the terms *sand*, *silt*, and *clay* were used in a very general fashion to mean different sizes of soil particles. The proper terms for different sizes (diameters) of soil particles are soil separates.

Soil Separates

Soil scientists in the United States recognize seven distinct classes of soil separates. These classes are given in Table 4-1.

Table 4-1
Sizes of the Soil Separates Recognized by the U.S. Department of Agriculture[1]

Soil Separate	*Diameter Limits (Millimeters)*[2]
Very coarse sand	2.0 -1.0
Coarse sand	1.0 - .5
Medium sand	.5 - .25
Fine sand	.25- .10
Very fine sand	.10- .05
Silt	.05- .002
Clay	Less than .002

[1] Soil Survey Staff, *Soil Taxonomy*, p. 469.
[2] A millimeter = 0.03737 inches.

Although Table 4-1 contains seven different soil separates, five of these separates are sands. Thus, the term *sand* can mean any soil particle which has a diameter between 2.0 and .05 millimeters.

Soil particles are classified into the different soil separates, because as particles become finer they have different properties. As particles in soil become finer the total amount of soil particle surface area becomes very large. This can be illustrated by thinking of a 1-inch cube. This cube will have six sides each containing 1 square inch and, thus, a total surface area of 6 square inches. If we cut this cube through the center, the resulting two pieces will have a total surface area of 8 square inches. Each time this cube is cut, the resulting particles will be smaller but the total surface area of the pieces will be larger. The surface area of the particles in a soil is quite important in determining the water-holding capacity of the soil. This is because much of the water in soils is held as films on the surfaces of the individual soil particles. Thus, finer-textured soils will have larger amounts of surface area and will hold larger amounts of water.

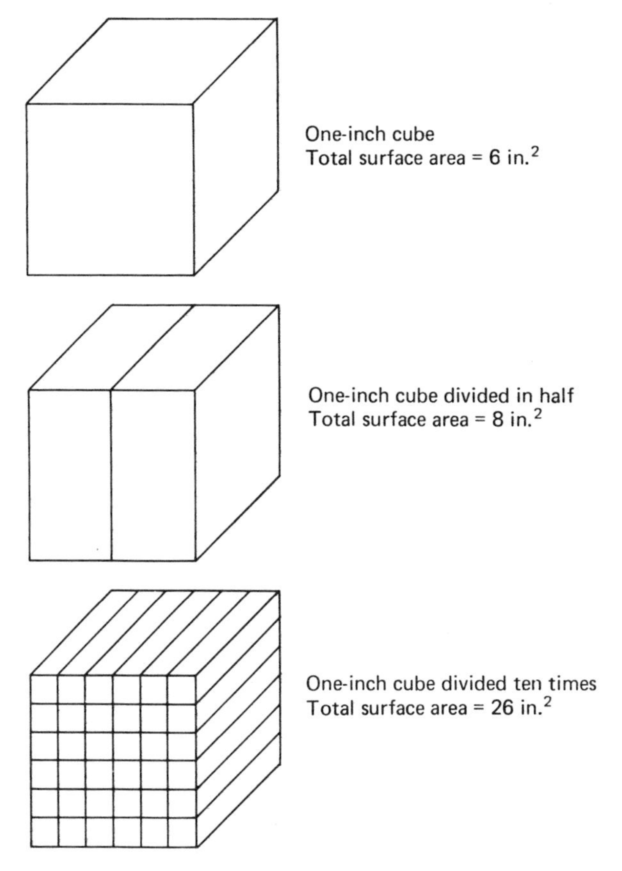

Figure 4-1. As a solid mass is subdivided, the resulting particles become smaller, but the total surface area increases. If this cube were divided 100 times, the surface area would be 206 inches squared.

Although the term *soil separate* refers to a specific size of soil particle and not to the composition of the particle, there are usually certain minerals which are dominant in the various separates. Sands are often composed of quartz and micas. Silts are commonly composed of a mixture of quartz and small mineral particles such as feldspars and micas. Clays are made up of secondary clay minerals which will be discussed in Chapter 6.

Soil Textural Classes

By this time it should be obvious that soils may be composed of all combinations of different amounts of the various soil separates. However, soil scientists group soils with similar amounts of sand, silt, and clay into groups called *soil textural classes*. There are four broad textural classes commonly referred to by soil managers. These textural classes are as follows:

1. Sands—soils containing more than 70 percent sand
2. Silts—soils containing more than 80 percent silt
3. Clays—soils containing more than 40 percent clay
4. Loams—intermediate mixture of sand, silt, and clay

In addition to these broad textural classes, soils are technically divided into the twelve textural classes presented in Figure 4-2. The textural triangle shown in Figure 4-2 is a graphic method of presenting the various classes of soils. The student should become acquainted with this diagram and be able to determine the textural class of a soil once he knows the amounts of sand, silt, and clay.

The textural class of a soil is determined in the laboratory by dispersing (separating) the soil particles in water, removing the sand with a sieve, and measuring the silt and clay by their rate of fall in water. This technique is called a *mechanical analysis* or *particle-size determination*. In the field, particle size is determined by feel. Although soils in the same textural class may feel slightly different depending on organic matter contents, coarseness of sand, type of clay, etc., guides were presented by Professor C. F. Shaw[1] in 1928 and are presently recognized by the National Cooperative Soil Survey as suggestive definitions.[2] The following guides are

---

[1] C. F. Shaw, *A Definition of Terms Used in Soil Literature* (Washington: First Internatl. Congress Soil Science Procedures and Papers 5, 1928), pp. 38-64.

[2] *Soil Taxonomy: A Basic System of Soil Classification for Making and Interpreting Soil Surveys*, Agriculture Handbook No. 436 (Washington: Soil Survey Staff, Soil Conservation Service, U.S. Department of Agriculture, Dec. 1975), p. 471.

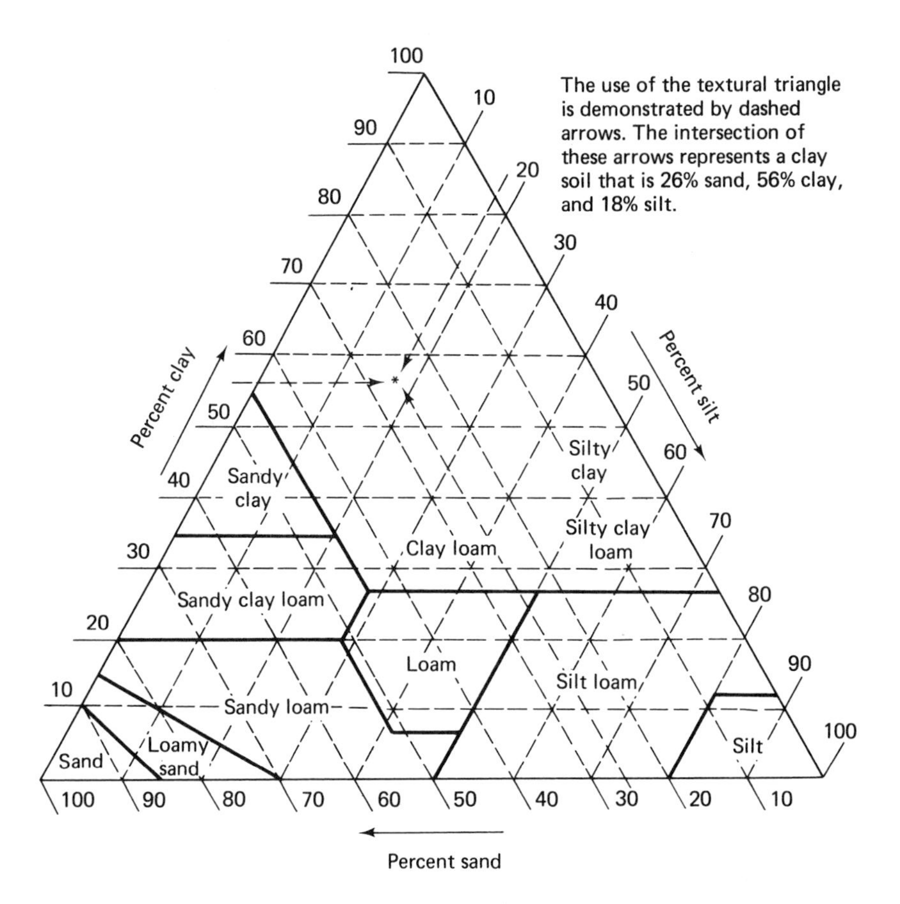

Figure 4-2. The textural triangle (a graphic representation of the amounts of sand, silt, and clay in the soil textural classes).

presented to assist the beginning student in determining the textural class of a soil. These guides are given for moist soils only. If the soil is dry, add water until the soil is moist but not wet to the point of free water being present.

Sand: Loose and single-grained with the individual grains readily visible; when squeezed a cast will form that will fall apart when the hand is opened.

Loamy Sand: Loose and single-grained with the individual grains visible; when squeezed a cast will form which will not fall apart when the hand is opened; the cast will break when handled.

Sandy Loam: Loose and single-grained with the individual grains visible; often sufficient clay present to give the appearance of coatings on the grains; when squeezed a cast will form that can be handled very carefully without breaking.

Loam: A relatively uniform mixture of sand, silt, and clay that may feel gritty but usually does not have complete visibility of sand grains; a cast formed by squeezing can be handled freely without breaking.

Silt Loam: Ranges from gritty to floury depending on size of the sand particles; usually visible sand grains appear coated when observed in the soil mass; a cast may be passed from hand to hand without breaking; soil will not ribbon and will give a broken appearance when pressed over the forefinger.

Silt: No visible sand grains; very smooth and floury feeling due to uniform particle sizes; may come close to forming a ribbon but will break in the process.

Sandy Clay Loam: Plastic soil; may have visible sand grains if pressed between thumb and forefinger; soil will form a ribbon which barely sustains its own weight.

Clay Loam: heavy, fairly uniform, plastic mass; few visible sand grains; will form a ribbon which barely sustains its own weight.

Silty Clay Loam: Heavy, plastic uniform mass that has a rough appearance when rubbed over the forefinger; forms a ribbon which will barely sustain its own weight.

Sandy Clay: Plastic to very plastic soil that will form a ribbon capable of sustaining its own weight; may appear to have a gritty feel or visible sand grains.

Silty Clay: Plastic soil that will form a ribbon capable of sustaining its own weight; will appear rough or broken when rubbed over the thumb and forefinger.

Clay: Plastic to very plastic soil that will appear greasy or sticky when rubbed over the forefinger; capable of forming a long ribbon which will support its own weight.

Beginning students should use these guides along with samples of known textural class to refine their ability to determine basic textural classes. The textural classes of sand, loamy sand, sandy loam, sandy clay loam, and sandy clay may also be defined in terms of the sizes of the sand particles present (i.e., fine sand, coarse loamy sand, etc.). The sand-size adjectives used range from very coarse to very fine as outlined in Table 4-1. The term *medium* is usually omitted when describing a sand class.

Whenever an appreciable number of particles greater than 2 mm are present, the textural class is often denoted with a descriptive adjective to denote the coarse particles. Examples of this are: gravelly clay loam, cobbly loam, stony clay, etc.

Soil Texture as Related to Soil Management and Land Use

Soil texture greatly affects the productivity and usability of a soil. For agricultural uses, coarse-textured soils such as sand and loamy sands cannot hold sufficient water or nutrients to produce top yields every year. Thus the soil manager must provide irrigation and frequent light fertilizer applications. Since water will infiltrate and percolate through sands very rapidly, they usually are not as susceptible to erosion as heavier soils which have less infiltration and greater runoff. Because sands can shift, they require spread footings and reinforced foundations when used as building sites. Waste water and septic tank effluent will percolate through sand quite easily. However, little filtration and purification may take place and ground water may be contaminated. On sandy sites, wells should be checked for nearby septic tank contamination. Sandy soils usually produce poor lawns which cannot stand heavy foot traffic and must be watered and fertilized frequently. When used for engineering purposes, sandy soils will not pack and thus are poor roadbed and fill materials.

Conversely, finer-textured soils such as clays, sandy clays, and silty clays may hold water and nutrients so tightly that they are unavailable for plant growth. They are also quite tight and may be difficult to till without becoming cloddy. Clays are quite susceptible to crusting and give rise to poor seedling emergence. Due to low infiltration rates, clayey soils may require terraces and strip cropping to prevent erosion. Provided the clays are a nonexpanding type (discussed in Chapter 6), clay soils provide good foundations for building sites and pack well for engineering uses.

When used as home sites, clay soils give problems in lawn establishment and may grow poorly due to compaction if much foot traffic is present. On low traffic areas clays can produce good lawns once the grass is established. Due to low infiltration rates on clayey soils, septic tanks seldom operate properly. The filter field often fails to absorb the effluent and it comes to the soil surface and causes a health hazard. Clay soils are excellent waste purifiers and seldom allow well contamination from septic tanks even when the filter field is in close proximity to the well.

Silty soils (silts and silt loams) give few agricultural problems but cause severe engineering problems because they shift and will not pack. This shifting occurs because silts are fairly uniform in size and have the stability of a stack of marbles. Whenever using silty soils for foundations, roadbeds, or steep banks, it would be wise to have competent advice on soils before beginning.

Intermediate-textured soils fall between the extremes found with sands, clays, and silts. They provide high agricultural productivity as well as giving few engineering, urban development, and waste management problems.

When considering soil texture in land use and soil management decisions, be sure to evaluate the total soil pedon rather than just the A1 horizon. Often sandy soils will be underlain by more clayey horizons. The total pedon will regulate water movement, waste effluent purification and movement, foundation stability, and plant rooting patterns. For example, the Norfolk soil of the South Atlantic coastal plain often has up to 50 centimeters of loamy sand A horizon which by itself would have all the problems of sandy soils. However, this loamy sand surface is underlain by a sandy clay loam B horizon containing 18 to 35 percent clay. This B horizon counteracts the problems of the sandy surface and gives an excellent soil for most uses. Likewise, there are many soils which have excellent A horizons underlain by either sands or clays. These soils can give rise to many land use and management problems.

SOIL STRUCTURE

Although soil structure has been recognized for many years, it is a physical property of soil which is not well understood. To envision soil structure consider soil particles (sand, silt, and clay) being cemented into clumps. These clumps are known as *aggregates* or *peds* to the soil scientist. The way these aggregates are arranged in the soil profile may form patterns. Thus we say that soil structure is the arrangement of soil particles into aggregates and the subsequent arrangement of these aggregates in the soil profile. The shapes of the individual soil aggregates or peds are usually quite distinct and easy to recognize. Technically the shape of the aggregates is called *structural type*. There are basically four types of structural units.

Types of Soil Structure

PLATY: Soil particles are formed into platelike aggregates that are arranged horizontally in the soil. This type of structure can actually reduce the penetration of air, water, and roots. It is

Figure 4-3. These blocky aggregates or peds are the building blocks of soil structure.

common in A2 horizons and seldom occurs in other horizons in the pedon.

PRISMLIKE: Soil particles are in prismlike peds arranged vertically in the soil. If the prismlike peds have flat or pointed tops, the type is called *prismatic*. This type is designated as *columnar* if the prismlike peds have rounded tops. Generally prismlike structure is found in the B horizons of midwestern soils and the sodium soils of the Great Plains. Usually this type of ped forms in soils containing montmorillonite clay (Chapter 6).

BLOCKY: These soil aggregates are divided into two types. If the angles on the sides of the peds approach 90° (right angles) the structure is called *angular blocky* or simply *blocky*. Blocky peds are very common in B2 horizons of many soils. If the angles on the sides of the peds are sharper than 90° the peds are referred to as *subangular blocky*. The subangular blocky units are often found in B1 and B3 horizons.

GRANULAR OR SPHEROIDAL: If the peds seem to be round and slightly porous, they are called *granular*. If they become very porous and almost resemble bread crumbs, the structural type is called *crumb*. The granular and crumb structures are very common in A1 and Ap horizons.

In addition to recognizing the types (shapes) of soil structure, soil scientists also recognize various sizes of structural units as structural class. The degree of development or distinctness is called the structural grade. The various types of structure are shown in Figure 4-4. For a detailed presentation of the various types and classes of structure, see Soil Taxonomy.[3]

In addition to the types of structure previously discussed, it should be noted that soils without peds present are referred to as *structureless*. If the soil particles are coarse and easily observable, the structure is called *single grain*. When the soil is a mass that does not break along structural planes and the individual grains are not present, the structure is called *massive*.

Formation of Soil Structure

As previously stated, soil structure and its formation are generally not well understood. The most accepted ideas regarding soil structure development are outlined as follows:

1. The soil contains the proper nutrients (calcium, hydrogen, and magnesium). The soil must definitely not contain much sodium or structure will be destroyed.

2. When the proper elements are present, soil particles will come together to form unstable aggregates. The factors which contribute to the bringing together of the soil particles to form these unstable aggregates are such things as: wetting and drying, freezing and thawing, root pressures, fungi which send out hairlike radials which encompass soil particles and pull them together, and soil tillage.

3. When these unstable aggregates are brought together they are then bound or cemented by some cementing agent such as: clay, organic matter, or iron and aluminum oxides.

Thus we see that the formation of soil structure is a multiphase process. It depends upon the soil being in the proper chemical balance and certain physical natural forces bringing the soil particles together into aggregates.

---

[3] *Soil Taxonomy*, p. 475.

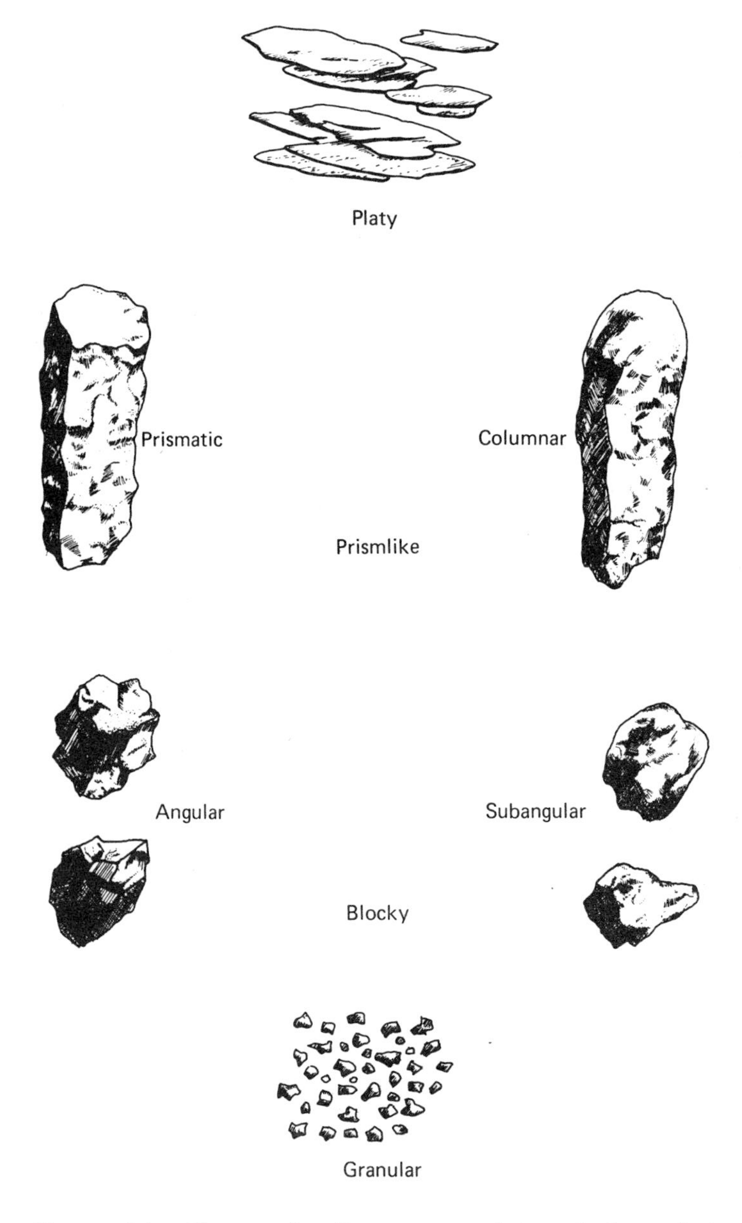

Figure 4-4. Types of soil structure. (adapted from *Soil Taxonomy: A Basic System of Soil Classification for Making and Interpreting Soil Surveys*, Agriculture Handbook No. 436, USDA, 1975, p. 475.)

These aggregates are then stabilized by natural cementing agents.

Importance of Soil Structure

Soil structure is very important in the topsoil because it increases permeability and thus cuts down runoff and decreases erosion. It enhances root growth by giving a more permeable soil through which roots can move. It gives better air relationships. By increasing the permeability to roots it may increase the effective water-holding capacity. That is, it may cause roots to explore a larger volume of soil, which makes them capable of extracting more water. In the subsoil structure can be even more important. It can enhance water movement and air permeability in heavy clay B horizons. Thus, it increases root penetration into these horizons. It also increases water percolation through these horizons and may cause better drainage of surface water and decrease erosion.

Ways to Improve Soil Structure

Because of the complex nature of soil structure, it is not possible for man to build it directly. However, there are several things which can be done to enhance the development of soil structure naturally. The following list of dos and don'ts will tend to help the soil build better structure.

1. Till soils only at the proper moisture contents. Never till when the soil is too wet. This causes the soil to become quite cloddy.

2. Add the proper amounts of lime. Without the proper amounts of lime present many beneficial soil organisms cannot grow and help to form soil structure.

3. Grow grasses and legumes whose top growth contributes organic matter and whose extensive root systems may form unstable aggregates. Grasses and legumes also furnish the organic matter to cement these unstable aggregates.

4. Grow legumes which give the soil more microorganisms and possibly enhance certain beneficial fungi to grow.

5. Turn under crop residues. Crop residues are probably the cheapest source of organic matter and nutrients which the farmer has at his disposal.

The previous procedures are common ways to enhance the structure of topsoils. There are no practical ways to greatly enhance subsoil structure. Subsoils will usually develop sufficient structure if treated properly. This means soil managers should avoid heavy machine compaction when the soil is wet. If possible it would be beneficial to grow deep-rooted legumes and grasses.

SOIL CONSISTENCE

Soil consistence refers to the forces of cohesion and adhesion exhibited by the soil, i.e., it is the degree of plasticity and stickiness of the soil. Soil consistence is determined by the type of clay in the soil. Do not confuse soil consistence with soil texture. Soil texture refers to the relative amounts of sand, silt, and clay in the soil; whereas soil consistence refers to the type of clay in the soil.

Determination of Soil Consistence

Usually, soil consistence is determined by observing soil feel when it is wet or moist. It can be determined in the laboratory by techniques called the *Atterberg limits*. Since these techniques are usually not available to the field soil manager, they need not be discussed. The following guidelines are given for determining field soil consistence.

PLASTICITY DETERMINATIONS: Determinations of plasticity can be made by pressing the soil between the thumb and the forefinger. If the soil is plastic it will be possible to press it into ribbons or various shapes. For the elementary student, imagine that molding clay is very plastic and sands, sandy loams, etc., are nonplastic.

STICKINESS DETERMINATIONS: Stickiness is usually determined by wetting the soil and seeing if it will stick to the finger. The degree of stickiness exhibited by a soil is related to the type of clay in the soil. Soils high in montmorillonite types of clay (Chapter 6) will be quite sticky; whereas soils high in kaolinite will not be nearly as sticky.

Importance and Use of Soil Consistence

Soil consistence coupled with soil texture tells us both the type and amount of clay present in a soil. If we know the properties exhibited by various types of clay and the amounts of these clays that are present in the soil, we have a sound basis for making management decisions. The following are some of the inferences that can be made about various soils.

PLASTIC SOILS: These soils will hold water quite well; however, they may or may not hold excessive amounts of nutrients, depending on the amount of kaolinite types of clay (Chapter 6) present.

STICKY SOILS: These soils are high in montmorillonite clay and will tend to have high water-holding capacities and high nutrient-holding capacities.

To thoroughly understand soil consistence, the student must have a background in the chemical properties of soils and the differences in the types of clay. This background will be presented in Chapter 6. At that time, soil consistence will then be reviewed in light of the new knowledge obtained by the student.

SOIL COLOR

Soil color has little actual effect on the soil; however, there are many things which we can tell about a soil by observing its color.

1. *Soil color and organic matter*: Soils high in organic matter are black or dark colored. The importance of organic matter will be pointed out later.

2. *Soil color as related to soil temperature*: Dark-colored soils absorb more heat; thus, they warm up more quickly in the spring and tend to exhibit higher soil temperatures.

3. *Soil color and parent materials*: Soils formed from mafic rocks will usually be darker in color and higher in nutrients than soils formed from felsic parent materials. Be careful, however, for this is not always true. There are cases where mafic rocks will develop into light-colored soils.

4. *Soil color and drainage*: Many soils have enough iron in them to turn red if they are oxidized (rusted). Soils which are well drained are red and yellow in color due to oxidized iron, while poorly drained soils have blue and gray colors due to reduced iron. Thus, we say

bright red and yellow soils are well drained while gray soils are poorly drained. Soil scientists actually classify soils into many different drainage classes. The four most common classes are well, moderately well, somewhat poorly, and poorly drained. These classes are distinguished in the field by the proportions of gray colors (mottles) in the soil profile. These gray mottles, or splotches of gray color, indicate that during some period of the year the soil is saturated with water for a prolonged period of time. Table 4-2 summarizes the occurrence of gray colors and the four major drainage classes.[4]

Table 4-2
Occurrence of Gray Colors in the
Four Major Soil Drainage Classes

Soil Drainage Class	*Occurrence of Gray Colors*
Well drained	No gray colors throughout the B horizon or to a depth of 60 inches
Moderately well drained	Gray colors in the B3 horizon or at a depth of 40 inches
Somewhat poorly drained	Gray colors in the upper B horizon starting at 20 inches
Poorly drained	Gray colors throughout the soil profile

SOIL PERMEABILITY

Soil permeability refers to the movement of air and water through soils. This movement of air and water is dependent on the amount and type of pore space present. For a soil to be permeable, it must contain pores which are continuous and large enough for water and air to pass through them. Just because a soil contains a large amount of pore space, it doesn't necessarily mean it is permeable. The pores could be discontinuous or very small.

Unlike the soil properties previously studied, soil permeability is not a property which can be readily seen or felt. It is also

[4] *Soil Survey Manual*, U.S. Department of Agriculture Handbook No. 18: Supplement (U.S. Department of Agriculture, May 1962), p. 155.

a property that is rather hard to measure quantitatively. However, there are several measurements which allow the soil manager to infer permeability. Two common measurements of pore space are total soil porosity and bulk density.

Soil Porosity or Pore Space

The pore space in a soil represents that part of a soil volume which can be occupied by air and/or water. It can be measured by saturating a soil and measuring the volume of water held. Although this gives a measure of total pore space, it does not tell what type of pore is present, i.e., large, small, continuous, discontinuous, etc.

Soil Bulk Density

The soil bulk density refers to the density of the soil in its natural state. It is the weight per unit volume of an undisturbed soil. Thus, soils high in pore space would have a low weight per unit volume (low bulk density). Likewise, soils low in pore space have high bulk densities.

Although both soil porosity and bulk density give a measure of pore space, they do not tell how fast water will move through the soil. To predict soil permeability we need to know the pore size and pore continuity. The measurement and calculation of these two factors are above the scope of this text. However, the soil manager can infer soil permeability by observing soil texture and structure.

Soil Permeability as it Relates to Soil Texture and Structure

Sands usually have high bulk densities (1.8 grams/cubic centimeter), low pore space (around 30 percent) but are quite permeable because the pores that are present are large and continuous. Conversely, clays have low bulk densities (1.2 to 1.3 grams/cubic centimeter), high pore space (around 50 percent), but are slowly permeable because the pores are small and often discontinuous.

Well-structured soils are more permeable than other soils of the same texture but lacking structure. This is due to larger pores around the soil aggregates increasing permeability.

Thus we see that soil permeability is related to pore size and pore continuity which in turn is related to soil texture and structure. If a soil manager wishes to change the permeability of a soil, he can change the texture or improve the structure. Textural changes are possible for small areas such as small lawns, flower gardens, etc. However, improved soil structure is the only economical means of changing large areas.

Soil Permeability and Soil Drainage

Common points of confusion with beginning students are the differences between the terms *soil drainage* and *soil permeability*. Technically soil drainage refers to the amount of oxidation which has taken place in the soil and permeability refers to water movement through soil. Thus, a sand could be very permeable but poorly drained if it is in a depression and a clay could be impermeable but well drained if on a ridge. Be careful not to confuse these terms.

OVERALL SOIL PEDON CHARACTERISTICS

With an understanding of soil physical properties in mind, we can now discuss the overall soil pedon. When a soil manager evaluates a soil he must not only be aware of the properties of the individual horizons but also of the arrangement of these horizons in the pedon. When evaluating the pedon, first observe the topsoil (total A horizon, i.e., A1 or Ap, A2, and A3 horizons). The topsoil is where tillage, root growth, and fertilization take place. Ask this question: "Does the topsoil have the proper texture, structure, and permeability for the use under consideration?" If not, either modify the soil or select another site.

Second, observe the subsoil or B and C horizons of the pedon. Ask yourself, "Will this subsoil allow proper root and water penetration for crop production or waste disposal?" or "Will this B horizon support buildings and roads if this is the intended use?" If these questions have to be answered no, select another site.

SUMMARY

Soil physical properties are those soil properties that contribute to the appearance and feel of a soil. They consist of soil texture, structure, consistence, color, permeability and overall pedon characteristics. Soil texture is determined by the relative amounts of sand, silt, and clay (soil separates) in the soil. Soils with similar amounts of each soil separate fall in the same soil textural class. Soil texture is quite important in soil management and land use because it affects water movement and percolation, which in turn affect plant growth, erosion, and waste management. Soil texture also directly controls engineering considerations such as foundation and roadbed stability.

Soil structure is the aggregation of the soil particles into peds. This aggregation, though not well understood, is very important in controlling root growth and water movement.

The consistence of a soil is controlled by the type of clay present. It is determined by the plasticity and stickiness of the wetted soil mass. The consistence of a soil reflects the nutrient- and moisture-holding capacities of a soil. Soil consistence will be considered again in Chapter 6.

There are many things that can be determined by observing soil color including organic matter contents, parent materials, and soil drainage. This is true even though soil color within itself has little effect on the soil.

Soil permeability is the only soil physical property which cannot be readily seen or felt. It is controlled by the number of soil pores and their continuity. The soil porosity is regulated by the soil texture and soil structure.

Whenever using soil physical characteristics to infer the properties of a soil, always consider the total soil pedon. The relationship between the A, B and C horizons is quite important.

REVIEW QUESTIONS

1. Can you define the following?

soil texture
soil separates
soil textural class
soil structure
platy structure
prismlike structure
blocklike structure
granular structure
textural type, grade, and class
soil consistence
color mottles
soil permeability
soil porosity
soil bulk density

2. What are the guides for determining each soil textural class by feeling a moist soil?

3. What is the textural class of a soil containing 40 percent sand, 30 percent clay and 30 percent silt?

4. How is soil texture related to soil management and land use?

5. What soil horizons contain the following types of soil

structure: Blocky? Subangular blocky? Prismatic? Columnar? Platy? Granular?

6. Outline the process of soil structure formation.

7. How is soil structure important in soil management and land use?

8. How can soil structure be improved?

9. How is soil consistence important in soil management and land use?

10. What can you tell about a soil by observing its color?

11. How is soil permeability related to soil texture and structure?

12. What is the difference between soil permeability and soil drainage?

REFERENCES

Baver, L. D. *Soil Physics* 3rd ed. New York: John Wiley, 1956, pp. 48-223.

Berger, Kermit C. *Introductory Soils*. New York: Macmillan Co., 1965, pp. 54-63.

Brady, Nyle C. *The Nature and Properties of Soils*. 8th ed. New York: Macmillan Co., 1974, pp. 40-66.

Donahue, Roy L., Shickluna, John C., and Robertson, Lynn S. *Soils: An Introduction to Soils and Plant Growth*. 3rd ed. Englewood Cliffs, N.J.: Prentice-Hall, 1971, pp. 30-51.

Foth, H. D., and Turk, L. M. *Fundamentals of Soil Science*. 5th ed. New York: John Wiley, 1972, pp. 27-62.

Soil Taxonomy: a Basic System of Soil Classification for Making and Interpreting Soil Surveys. Soil Survey Staff, Soil Conservation Service, U.S. Department of Agriculture. U.S. Department of Agriculture Handbook No. 436, Dec. 1975, pp. 463-79.

Soil Survey Manual, U.S. Department of Agriculture Handbook No. 18: Supplement. U.S. Department of Agriculture, May 1962, pp. 189-238.

5

Soil Water

"Why is the soil storing all that rain water?"
"Because it has a crop to grow."

One of the most important functions a soil performs for a growing crop is catching water during periods of rainfall and storing it for the plants to use at a later time. If a soil is not capable of performing this task, it is of limited value for nonirrigated crop production. The processes of soil water storage and movement can be very complex and highly mathematical. However, a general understanding of the processes is essential in order to make decisions on basic irrigation and land use problems. In this text we will give a very general and descriptive discussion of soil moisture so the student can visualize basic concepts of water movement and uptake. Mathematical discussions will be limited to the practical examples of soil moisture calculations presented in Appendix 5-1.

TYPES OF WATER ASSOCIATED WITH THE SOIL

To start our discussion of soil water let us look at the three basic types or forms of moisture which can occur in the soil. All of these moisture forms start as free water added to the soil by rainfall or irrigation. Their final form depends on the moisture level of the soil.

GRAVITATIONAL WATER: Free water that can move through the soil due to the forces of gravity is called gravitational water. This water is usually not used by plants because it rapidly moves out of the soil. This type of water can cause plants to wilt and die because it occupies necessary air space in the root zone.

CAPILLARY WATER: This is water that is in the soil pores or held very loosely around the soil particles. Most of this is the water which is available to plants for growth and transpiration.

HYGROSCOPIC WATER: This water forms very thin films around soil particles and is not usually available to plants. Some of this water is in soils even when they are air dry. Hygroscopic water is attracted very tightly to the soil.

Gravity is always acting to remove water from the soil by causing downward percolation. However, the force of gravity is counteracted by attractive forces between water molecules and soil particles as well as the attraction between water molecules themselves.[1] These attractive forces countering gravity give rise to the previously discussed forms of soil water.

FORCES ACTING ON SOIL WATER

Gravitational water moves under the force of gravity. About twenty-four to forty-eight hours after a heavy rain the gravitational water in the soil is removed by gravitational forces. Capillary water is held by cohesion (attraction) between water molecules. This force is not too strong yet it is sufficient to counteract gravity. Hygroscopic water is held by forces of adhesion between the soil particles and water molecules. It is held very tightly. The various forms of water are illustrated in the schematic diagram presented in Figure 5-1.

ENERGY RELATIONSHIPS OF SOIL WATER

When working with soil water we are usually interested in the strength with which water is held in the soil. A measure of how much force with which water is held by the soil is the bar. One bar equals about one atmosphere of pressure or is the pressure

---

[1] Due to their molecular structure, water molecules tend to act somewhat similarly to bar magnets with distinct negative and positive poles. The negative charge on soils may attract the positive pole of the water molecule. Additional layers of water may be held by the water molecules stacking or layering in a − + − + fashion.

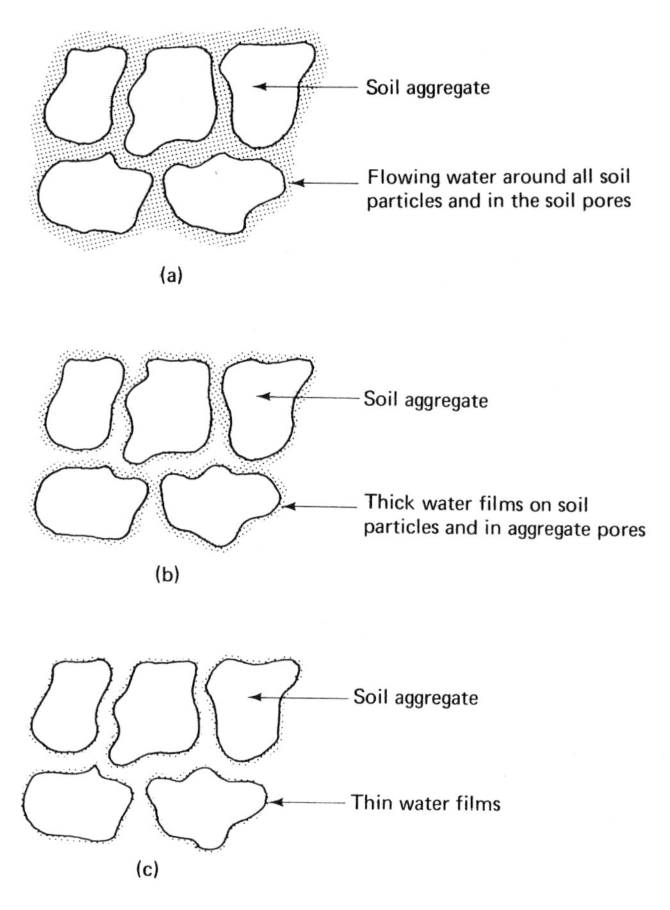

Figure 5-1. Forms of soil water: (a) gravitational water—excess water that fills the soil pores and flows due to gravitational force; (b) capillary water—thick water films held by the forces of cohesion between water molecules; (c) hygroscopic water—thin water films held by the forces of adhesion between the soil particles and water molecules.

exerted at sea level by the atmosphere.[2] In the laboratory we can produce pressures ranging from zero pressure to sixteen bars by using simple compressed gas techniques. These pressures are useful in evaluating irrigation techniques because of the following:

---

[2] One bar equals the pressure applied by a water column 1 centimeter in area and 1,000 centimeters in height. Whereas one atmosphere equals the same column area but 1,033 centimeters in height.

1. From zero to about one-third bar pressure all gravitational water drains.

2. From one-third bar to fifteen bars capillary water drains.

3. From fifteen bars upward we have hygroscopic water.

In this text we are not too concerned with energy relationships, except as they are used in irrigation. Remember we said capillary water is available to plants and that this water is held so that it takes about one-third to fifteen bars to remove it. The exact amount of pressure to remove plant available moisture is dependent on many factors; thus the one-third to fifteen bars is only an estimate. In general irrigation work we usually strive to hold the soil moisture between one-third and one bar pressure. For special crops less than one-half bar may be used and for heavier soils it may be possible to let them dry to as much as three to four bars.

SOIL MOISTURE CONSTANTS

When working with soil moisture there are several moisture constants between soil saturation and air dry. These constants are quite important in irrigation work and in an understanding of soil moisture. These constants are as follows:

1. *Soil saturation* When the soil contains all the water it can hold without water standing on top of the soil. Soil reaches this point during and immediately after a rain.

2. *Field capacity* Moisture content of the soil after gravity has removed all the water it can. This is usually one to three days after a rain depending on the soil texture and structure.

3. *Wilting point or percentage* That percentage of soil moisture at which plants cannot obtain enough moisture to keep from wilting. Depending on soil type, this moisture content is usually around fifteen bars of pressure.

4. *Hygroscopic percentage* The moisture content of the soil when it is about air dry.

5. *Oven dryness* This condition exists when the soil has been heated at 105°C for at least twelve hours and

all soil moisture has been removed. This point is not important in plant growth but it serves as a reference point for most moisture calculations.

FORMS OF SOIL WATER AND PLANT GROWTH

By this time the student should be acquainted with the various forms of soil water. Figure 5-2 relates these various forms of water to plant growth. From Figure 5-2 it should be evident that only part of the capillary water furnishes all of the plant-available moisture supplied by the soil. It should also be evident that the only plant-available moisture in the soil is that moisture which occurs between field capacity and the wilting point.

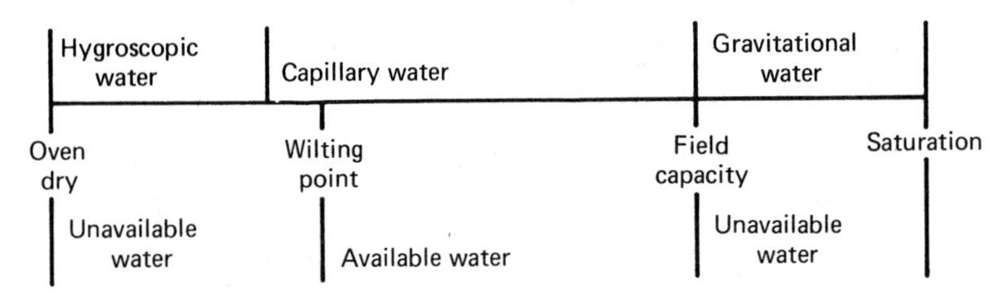

Figure 5-2. Total soil moisture and its relationship to plant available moisture.

WATER REQUIREMENTS OF CROPS

The amount of water required to produce a pound of dry plant material is called the transpiration ratio. This ratio varies for different crops and different climatic conditions. In arid regions where humidity is low, the transpiration ratio is high. Whereas, in humid regions, the ratio is somewhat reduced. However, the ratio generally ranges from 250 to 750 pounds of water per pound of dry matter produced. From these figures it can readily be seen that a tremendous amount of moisture is required to produce a twenty-ton crop of corn silage.

SOIL MOISTURE CALCULATIONS

With the previous discussion of soil moisture in mind we are now ready to calculate soil moisture values. There are various ways to determine soil moisture. However, they all depend on the percentage moisture on an oven-dry basis as a standard. This calculation is made as follows:

$$\text{Percentage of soil moisture} = \frac{\text{weight of soil moisture}}{\text{dry soil weight}} \times 100$$

In this equation the dry soil weight refers to the oven-dry soil weight unless otherwise specified. The weight of the soil moisture is the soil moisture lost in drying. For example, if a moist soil weighed 500 grams and the same sample oven dried weighed 400 grams, the weight of the soil moisture would be 500 grams minus 400 grams or 100 grams of soil moisture. Likewise the percent moisture would be:

$$\frac{100 \text{ grams}}{400 \text{ grams}} \times 100 = 25 \text{ percent}$$

The examples in Appendix 5-1 should further clarify the use of this formula and other soil moisture calculations.

FACTORS AFFECTING SOIL WATER-HOLDING CAPACITY

We are quite interested in these factors since they are the way we can influence and control the soil's moisture-holding capacity.

SOIL TEXTURE: Remember soil water occurs as films around soil particles; therefore, the more clay there is in a soil, the more water it will hold. This does not necessarily mean there will be more available water. The graph in Figure 5-3 should demonstrate this point.

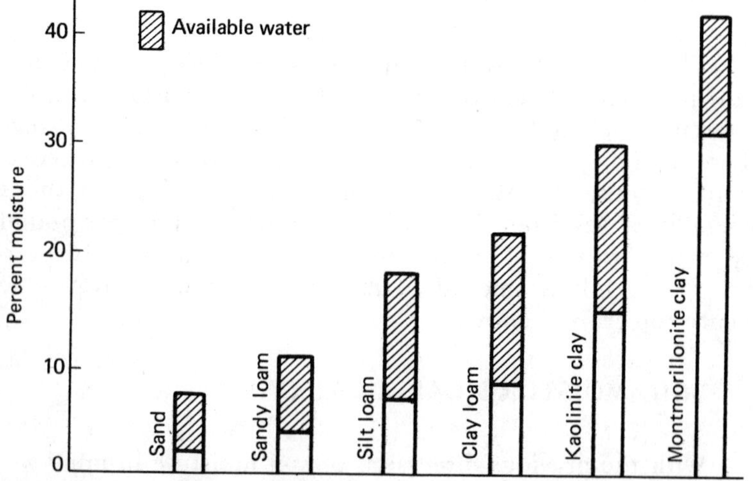

Figure 5-3. Relationship among various soil textures and moisture-holding capacity. Each bar represents the water between oven dryness and field capacity.

SOIL CONSISTENCE: The type of clay present greatly affects water holding capacities of soils. Kaolinite clay is usually much coarser than montmorillonite and thus holds less water. This is also substantiated in Figure 5-3.

Although there are very few things we can do to modify soil texture and consistence, we can be aware of their effects and plan our land use accordingly.

SOIL STRUCTURE: Soil structure does not directly affect the soil's ability to hold moisture. However, it does make the soil more permeable and allows gravitational water to drain more quickly. It also may allow plants to root deeper and exploit moisture from more of the soil profile. When this happens we can say it increases the "effective" water-holding capacity. Soil structure is often the soil manager's only tool in increasing effective available moisture.

ORGANIC MATTER: Organic matter affects the soil's moisture-holding capacity in two ways:

1. It enhances soil structure and aggregation.

2. It greatly increases the moisture-holding capacity if present in large amounts. When large amounts of organic matter are present the soil may have a water-holding capacity of greater than 100 percent. This is because the organic matter is very light and swells as it becomes wet. Thus, greater than 100 percent moisture still may not be a large amount of actual water.

MOVEMENT OF WATER IN SOILS

For the soil manager to make wise use of soil water and apply the proper irrigation, it is necessary to understand the movement of the various forms of soil water.

Gravitational Water

As previously stated gravitational water is free water. This means it simply flows through or off the soil. Gravitational water is responsible for nutrient leaching and soil erosion. Thus, on soils with low water-holding capacities it is essential that nutrients be applied when the plants need them. Otherwise, they may be leached below the plant root zone. This is quite important on coarse sandy soils. When applying irrigation water the manager should always regulate flow rates and amounts of water so that leaching and erosion do not occur.

Capillary Water

Unlike gravitational water, capillary water does not flow through the soil. It occurs as water films around the individual soil particles and simply moves to the points of highest tension. To envision this, think of a soil particle having a thick water film around it held at one-third bar of pressure. Now envision a root hair extracting moisture from this water film. As the root hair removes water from the film on one side of the particle, the tension holding the water at the point of contact will increase to a tension greater than one-third bar. As this happens, the water from the film on the other side of the particle will move around to the point of root contact and try to bring the overall water film tension into equilibrium. This water movement is demonstrated schematically in Figure 5-4.

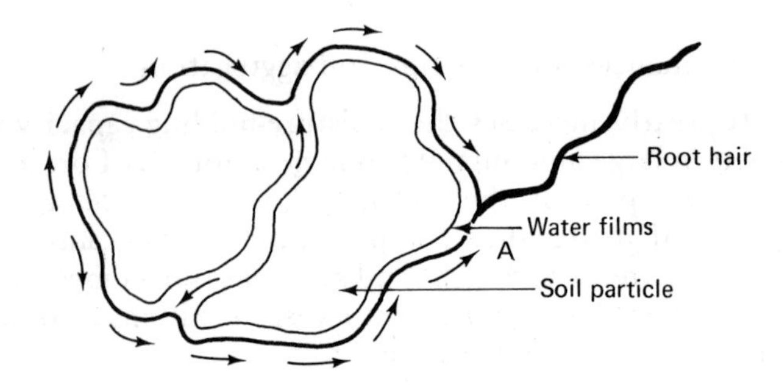

Figure 5-4. The movement of capillary water: as the root hair removes water at point A, water moves to point A as indicated by the arrows.

Plant roots will continue to remove capillary water until the attraction between the water molecules and the soil particles becomes greater than attractive forces of the roots. When this happens the permanent wilting point is reached. Prior to reaching the permanent wilting point, temporary wilting may occur during hot sunny days. This temporary wilting occurs when the water films get quite thin and the capillary movement slows down.

From the previous discussion it should be obvious that the goal of irrigation is to increase the thickness of the capillary films in the plant root zone. If too much water is applied, gravitational water will occur and cause leaching. If too little water is applied, the water films will not be thick enough to prevent temporary wilting.

Hygroscopic Water

Hygroscopic water moves under gravity's influence as a vapor. It has long been assumed that water vapor movement was not important in plant growth. However, some recent studies have shown that this assumption may not be correct. In either case, the movement of hygroscopic water is quite complex and beyond the scope of this text.

CAPILLARY RISE OF WATER

A discussion of soil water would not be complete without mentioning the capillary rise of water. Capillary rise occurs when water from a water table moves upward through the soil profile against the force of gravity. It is caused by the attraction between the soil particles and water molecules being greater than the force of gravity. The height to which the water will rise depends on the size of the soil pores. If the pores are large the height will be small, because the forces of gravity will soon counteract the molecular forces. If the pores are very small then there will be a greater rise because the molecular forces will be large relative to the amount of water in the pores.

On well-drained upland soils, capillary rise will not be an important source of soil moisture for plant growth. However, recent investigations have shown that on poorly drained soils with water tables within 40 inches of the soil surface, capillary rise can supply much moisture. Because of these findings much work is being done on water table control as a source of subsurface irrigation.

SOIL WATER MANAGEMENT

The goal of anyone growing crops should be to utilize soil moisture as efficiently as possible. This is quite important because rainfall is often scarce and irrigation is very expensive. The key to efficient water utilization is to reduce water losses and increase plant use efficiency.

Ways to Reduce Water Loss

Because the soil manager cannot control rainfall amounts and distribution, he does not have complete control of soil moisture. However, there are certain things which can be done to reduce soil moisture losses. Table 5-1 lists the various ways water can be lost and gives possible control measures.

Table 5-1
Methods of Controlling Soil Moisture Losses

Type of Moisture Loss	*Control Measure*
Runoff	Use terraces, strip crop, add grasses to the rotation, increase infiltration by enhancing soil structure
Evaporation	
Soil surface	Space plants to give a complete soil cover
Plant surfaces	None
Transpiration	Remove weeds and grasses from growing crop
Percolation	Increase soil moisture-holding capacity on small areas, add organic matter from crop residues

Increasing Plant Efficiency

There are certain practices which will increase the plant efficiency of soil water use. These are as follows:

1. Plant crops at the proper time to utilize seasonal rainfall.

2. Plant varieties adapted to meet water availability conditions.

3. Use plant-spacing arrangements which will fully utilize soil moisture.

4. Avoid compaction causing plow pans which will limit rooting depths.

5. Arrange crops on the landscape to take advantage of varying soil moisture-holding capacities (for example, coastal bermuda on deep sands, corn on poorly drained wet soils, tobacco on well-drained uplands, etc.).

SUMMARY

There are three types of soil water; gravitational water which moves through the soil by the forces of gravity; capillary water which is held loosely around soil particles and in soil pores, and hygroscopic water which is very tightly held around soil particles. Gravitational water usually drains from the soil twenty-four to forty-eight hours after a rain. Capillary water counteracts gravity and requires roughly one-third to fifteen bars of pressure to extract it from the soil. Hygroscopic water is that water which is in the soil at air dryness or held by the soil at tensions above fifteen bars.

When dealing with soil moisture we are interested in several key soil moisture contents. These are soil saturation, field capacity, wilting point, hygroscopic percentage, and oven dryness. Each of these points represents an important characteristic for a given soil. For example, the only soil water that is available for plant growth is a portion of the capillary water which lies between field capacity and the wilting point.

Soil water-holding capacity is affected by soil texture, soil consistence, soil structure, and soil organic matter contents. The soil manager can use these factors along with a knowledge of soil water movement to assist in developing soil water management practices which will reduce soil water losses and enhance plant efficiency.

REVIEW QUESTIONS

1. Define the various forms of soil water.
2. Define soil saturation, field capacity, and wilting point.
3. Describe the forces that act on the various forms of soil moisture.
4. What is available water? When does it occur in the soil?
5. Do you understand the soil moisture problems in Appendix 5-1?
6. Discuss the factors affecting the soil's moisture-holding capacity.
7. How do the various forms of water move in the soil?
8. What is meant by capillary rise of water?
9. How can water losses be reduced?
10. What management practices can increase the efficiency of water use?

REFERENCES

Baver, L. D. *Soil Physics*. 3rd ed. New York: John Wiley, 1956, pp. 224-303.

Berger, Kermit C. *Introductory Soils*. New York: Macmillan Co., 1965, pp. 64-77.

Brady, Nyle C. *The Nature and Properties of Soils*. 8th ed. New York: Macmillan Co., 1974, pp. 164-227.

Donahue, Roy L., Shickluna, John C., and Robertson, Lynn S. *Soils: an Introduction to Soils and Plant Growth*. 3rd ed. Englewood Cliffs, N.J.: Prentice-Hall, 1971, pp. 207-21, 337, 346-65, 379-403.

Foth, H. D., and Turk, L. M. *Fundamentals of Soil Science*. 5th ed. New York: John Wiley, 1972, pp. 63-88, 395-412.

Appendix 5-1

Soil Moisture Calculations

The following four soil moisture problems are presented to demonstrate the use of equations to caluclate soil moisture, as well as to assist the student in envisioning the forms of soil water. After studying these examples it should be possible to work the problems at the end of this Appendix.

Example 1

A soil weighs 220 grams when moist. After drying at 105°C for 12 hours it weighs 200 grams. What is the percent of moisture?

Solution:

a. wt. of soil moisture $= 220\text{g} - 200\text{g} = 20\text{g}$

b. percent moisture $= \frac{20\text{g}}{200\text{g}} \times 100 = 10\%$

The percent soil moisture formula also applies to many other soil moisture problems concerning the soil moisture constants and available water. The following example should demonstrate this.

Example 2

After a large soaking rain, a soil was sampled as it dried. The following weights were observed.

a. Immediately after the rain–300 grams

b. Two days after the rain–270 grams

c. Five days after the rain–250 grams

d. When plants growing on the soil wilt–220 grams

e. When the soil is air dry–210 grams

f. When the soil is oven dry–200 grams

Find the percent moisture at: saturation, field capacity, five days, the wilting point, and air dryness. Also, find the total available water, the available water at five days, and gravitational water.

Solution:

For the beginning student, the best approach to a problem such as this one is to first construct a chart such as the one presented in Figure 5-2. Next impose on this chart the soil weights at various moisture contents.

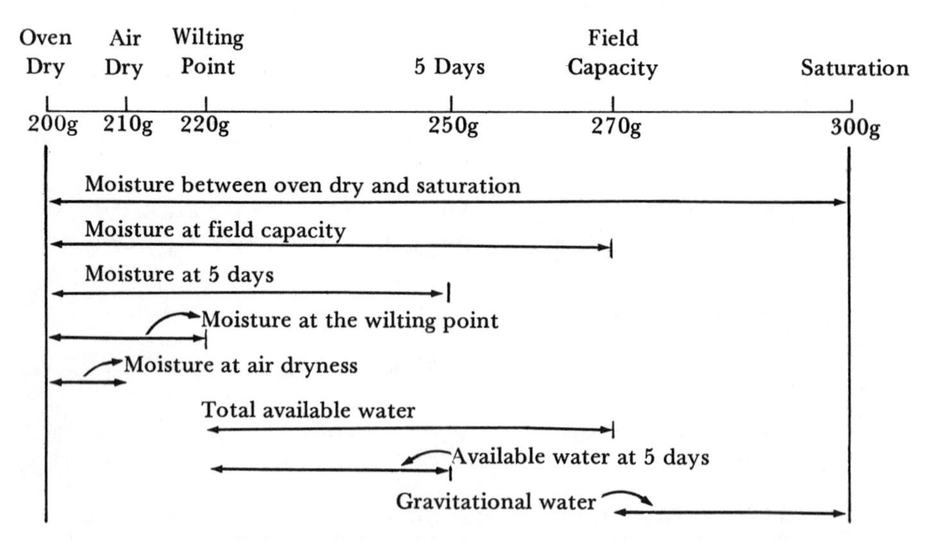

This chart will give a pictorial diagram of the moisture contents we are dealing with. Now make the calculations by using the percent moisture formula.

a. percent moisture at saturation

(1) weight of soil moisture = 300g − 200g = 100g

(2) percent moisture = $\frac{100g}{200g}$ X 100 = 50%

b. percent moisture at field capacity

(1) weight of soil moisture = 270g − 200g = 70g

(2) percent moisture = $\frac{70g}{200g}$ X 100 = 35%

c. percent moisture at 5 days

(1) weight of soil moisture = 250g − 200g = 50g

(2) percent moisture = $\frac{50g}{200g}$ X 100 = 25%

d. percent moisture at the wilting point

(1) weight of soil moisture = 220g − 200g = 20g

(2) percent moisture = $\frac{20g}{200g}$ X 100 = 10%

e. percent moisture at air dryness

(1) weight of soil moisture = 210g − 200g = 10g

(2) percent moisture = $\frac{10g}{200g}$ X 100 = 5%

f. percent total available water

(1) weight of soil moisture = 270g − 220g = 50g

(2) percent moisture = $\frac{50g}{200g}$ X 100 = 25%

g. percent available water at 5 days

(1) weight of soil moisture = 250g − 220g = 30g

(2) percent moisture = $\frac{30g}{200g}$ X 100 = 15%

h. percent gravitational water

(1) weight of soil moisture = 300g − 270g = 30g

(2) percent moisture = $\frac{300g}{200g}$ X 100 = 15%

Because the percent moisture at saturation, field capacity, 5 days, and the wilting point had been calculated in steps a through d, answers f, g, and h could have been obtained by subtraction, e.g., percent total available moisture = percent moisture at field capacity minus percent moisture at the wilting point = 35% – 10% = 25%, etc.

The previous example may also be used to clarify a point of confusion which often plagues beginning students. Suppose for some reason the oven-dry weight of the soil was not found. Then, all calculations would have to be based on the air dry weight. This does not complicate the problem as long as the student knows which weight to use. Employing the data of the previous example, we find that the total available soil moisture on an *air dry basis* would be:

(1) weight of soil moisture = 270g – 220g = 50g

(2) percent moisture $= \frac{50\text{g}}{210\text{g}} \times 100 = 23.8\%$

The calculations of the previous two examples were to prepare for more complicated moisture problems. The following example illustrates the use of soil moisture data to calculate some practical irrigation information.

Example 3

Suppose you were given the following information:

a. An acre of soil 6 inches deep weighs 2 million pounds.

b. One cubic foot of water weighs 62.4 pounds.

c. The transpiration ratio of corn is 400 pounds of water per pound of dry matter.

d. Corn grain weighs 56 pounds per bushel.

e. About 6,000 pounds of stalks and cobs are in a 100-bushel corn crop.

f. There are 43,560 square feet in an acre.

Find how many inches of water are needed to produce 100 bushels per acre.

Solution:

a. 100 bu. X 56 lb./bu. = 5,600 lb.
stalks, leaves, and cobs = 6,000 lb.
total dry matter produced = 11,600 lb.

b. 11,600 lb. of dry matter X 400 lb. of water/lb. of dry matter = 4,640,000 lb. of water needed.

c. If a cubic foot. of water weights 62.4 lb., then a square foot of water 1-inch deep must weight 62.4 lb./12″ or 5.2 lb./sq.ft. 1-inch deep, and 5.2 lb./sq. ft. inch X 43,560 sq. ft./A = 226,512 lb. of water/acre inch. (226,512 lb = the wt. of 1-inch of water over an acre).

d. Therefore $\frac{4{,}640{,}000 \text{ lb.}}{226{,}512 \text{ lb./A in.}} = 20.5''$ of water needed.

Example 4

Suppose the soil in Example 3 had a wilting point of 10 percent and a field capacity of 25 percent. Assume the growing corn could use moisture to a depth of 2 feet. If a saturating rain fell after the corn was planted, how much water would the soil be able to supply to the crop?

Solution:

a. If 6 inches of soil over an acre weighs 2 million pounds, 2 ft. must weigh 8 million pounds.

b. If the field capacity is 25 percent and the wilting point 10 percent, then the available water must equal 15 percent.

c. Therefore: 8,000,000 lb. X .15 = 1,200,000 lb. of available water, which, when divided by 226, 512 lb/A inch, is 5.3 inches of available water.

Summary of Moisture Calculations

The previous four examples were given to demonstrate that soil moisture calculations are detailed but quite simple, provided the student understands the properties of soil moisture. They were also given to show the large amounts of water necessary for crop production.

The following soil moisture problems are presented in order that the student can practice soil moisture calculations.

1. Given: Wet soil weighs 600 grams. When oven dry the soil weighs 500 grams. Calculate the percent moisture and pounds of water per acre furrow slice (assume acre-furrow of soil weighs 2 million pounds).

2. Assume the above soil was at field capacity when sampled

and that the wilting point for this soil was 5 percent. Calculate the percent available water.

3. If an acre-inch of water weighs 226,500 pounds, to how many inches of rain would the available water in question 2 be equivalent?

4. Given: Soil has a wilting point of 11 percent and a field capacity of 26 percent. An acre-furrow slice = 6 inches. How many inches of water would be required to penetrate the soil to a depth of 2 feet if it is at the wilting point when the rain starts?

5. Corn is planted on the soil in problem 4 when the moisture content is 21 percent. If 13 inches of rain falls during the growing season and 75 percent of this is utilized by the plant, how many pounds of dry matter can be produced per acre? Take the transpiration ratio of corn to be 350. Assume that the top 2 feet of soil are exploited by the root system and all available water is used.

6. Assume that a farmer with limited acreage wants to make a minimum of 6 tons per acre of clover hay. He is growing the hay on a soil with a field capacity of 20 percent and a wilting point of 8 percent. The crop exploits 2 feet of soil and uses all of the available water in this area. If clover has a transpiration ratio of 550 and it only rains 8 inches during the season, how much irrigation water will have to be added? (in acre-inches or pounds per acre).

6

Soil Chemical Properties

"What are soil's chemical properties?"

"Colloidal fraction of the soil."

"Soil properties that require a chemical reaction to be effective, and percent base saturation."

"Cation exchange capacity—CEC for short."

In previous chapters, we have been interested primarily in how soils are formed and in some of their physical properties. We are now ready to look at some of the chemical properties possessed by soils. These chemical properties give soils their ability to hold nutrients and create a desirable chemical environment for plant growth. In Appendix 6-1, at the end of this chapter, ten simple chemical definitions are presented. It is the student's responsibility to become acquainted with these definitions as they will be used during the remainder of this chapter and in those which follow.

In Chapter 4 it was observed that soil particles could be divided into sands, silts, and clays. When studying soil chemical properties we are concerned mainly with the soil clays and organic matter. More specifically we will be concerned with particles known as soil colloids. A soil colloid is a soil particle which is small enough to stay suspended in water. This usually happens when the soil particle is less than 0.002 mm in diameter. For a particle to stay suspended it usually carries an electrical charge. On most soil colloids, this charge is negative.

TYPES OF SOIL COLLOIDS

There are numerous types of soil colloids. In this text only the most important will be discussed. Most of the additional references for this chapter contain discussions of other soil colloids.

Silicate or Soil Clay Colloids

In general, there are two distinct types of soil clays with several subtypes within each type. These are the 1:1 clays (kaolinite) and the 2:1 clays (montmorillonite).

KAOLINITE OR 1:1 TYPE CLAYS: Schematically, 1:1 clays are depicted in Figure 6-1. The negative charges on the 1:1 clays come from broken chemical bonds along the edges of the clay particles. These broken chemical bonds occur when the kaolinite is broken into small pieces during weathering. The oxygen-hydrogen linkage holds the 1:1 plates apart at a fixed distance. There are two major 1:1 clay minerals present in the soil. They are known as kaolinite and halloysite.[1] For most practical purposes they act quite similarly in the soil.

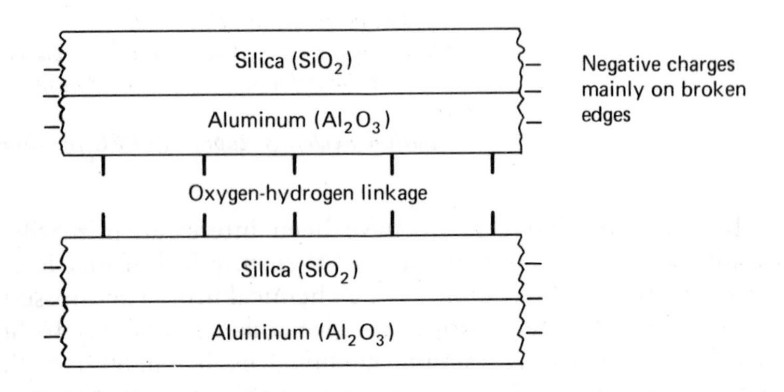

Figure 6-1. Schematic diagram of a 1:1 clay mineral.

MONTMORILLONITE OR 2:1 TYPE CLAYS: The 2:1 type clays are shown schematically in Figure 6-2. The negative charges on 2:1 clays come both from broken bonds on the particle edges and from molecular arrangements within the aluminum layer. Often up to 20 percent iron and/or magnesium will substitute for aluminum in the aluminum layer. This means an ion with a valence of plus 2 is substituting for aluminum with a plus 3 valence. The effect is a need for additional positive charges. Thus, the overall net effect is for the colloid to carry a negative charge.

---

[1] Kaolinite is the name given to the general category of 1:1 clay minerals. However, it may also be used as a specific name for a 1:1 mineral. For the beginning student the distinction is of little importance as long as kaolinite is associated with 1:1 clay minerals.

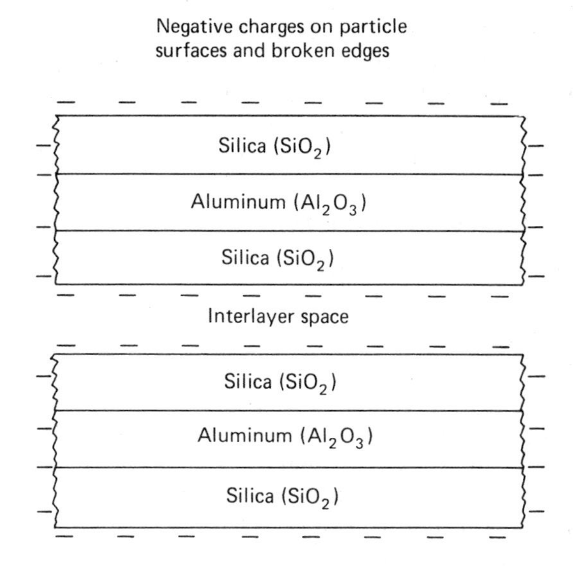

Figure 6-2. Schematic diagram of a 2:1 clay mineral.

There are several types of 2:1 clays. The differences in these clays are the materials in the interlayer space and the molecular arrangements in the aluminum layer. For our purposes we will only be interested in four 2:1 types of clays. These are montmorillonite, illite or clay mica, vermiculite, and vermiculite-chlorite materials.[2]

Montmorillonite Contains water in the interlayer space. This allows it to expand when wet and shrink when dry. Trapped ions may also be found in the interlayer.

Illite or Clay Mica Potassium is part of the clay structure. Illite has a different molecular arrangement in the aluminum layer which reduces the total negative charge. In soils where this type of clay is present, high levels of potassium may be found.

Vermiculite This clay mineral is similar to illite except magnesium is present in the mineral structure. The negative charge is greater than montmorillonite.

Vermiculite-Chlorite Materials These are materials that are somewhere between vermiculite, a 2:1 mineral, and chlorite, a 2:2 mineral. In these materials the interlayer space is being filled with

---

[2] Montmorillonite is the name given to a general class of 2:1 clays, as well as the name for a specific 2:1 clay within the general 2:1 class. These clays are often called smectites.

aluminum to give two silica and two aluminum layers or a 2:2 mineral called chlorite. The properties of these materials are quite variable depending on the amount of aluminum in the interlayer space. These materials are discussed because they are quite extensive in the Southeast. The aluminum interlayer is quite important because it can contribute to soil acidity and fertility problems.

Hydrous Oxides or Hydroxy Iron and Aluminum

These form a second general type of inorganic colloid. They are important because this type of colloidal matter is often dominant in the soils of the tropics and semitropics, as well as being quite common in the southeastern United States. These oxides of iron and aluminum occur as coatings on soil particles and/or discrete clumps in the soil. These small particles may carry negative charges and act as a point around which cations are attracted. The cation adsorption capacity is less than for kaolinite. Their charge is quite dependent on soil acidity levels. As the soil acidity increases, the charges on these colloids usually decrease in number.

Most hydrous oxides are not as sticky, plastic, and cohesive as are the silicates. This helps to explain why soils dominated by these colloids possess superior physical properties than those strongly influenced by the silicate clays previously described.

These colloids often possess positive as well as negative charges and may contribute to the adsorption of negatively charged fertilizers. An example of this type of reaction is shown below involving phosphate ions being adsorbed on an aluminum hydroxide surface.

$$\mathrm{Al}\begin{cases}\mathrm{OH}\\ \mathrm{OH}\\ \mathrm{OH}\end{cases} + (\mathrm{H_2PO_4})^- \longrightarrow \mathrm{Al}\begin{cases}\mathrm{OH}\\ \mathrm{OH}\\ \mathrm{H_2PO_4}\end{cases} + \mathrm{OH^-}$$

This type of reaction is a component of "phosphate fixation" (Chapter 11) and accounts for the tie-up of a large part of fertilizer phosphorus when applied to highly weathered soils.

Organic Colloids

Organic matter is of a colloidal nature when it decomposes. Unlike clay minerals, organic colloids seldom have a distinct structure. These colloids simply occur as coatings around soil particles in the Ap horizons of mineral soils or as the main colloid in organic soils. Organic colloids become quite important in the Tidewater region of the Atlantic Coastal Plain and the organic areas of glaciated northern and midwestern United States.

CATION EXCHANGE

Because of the negative charge on soil colloids, they are able to attract positively charged cations. Schematically, this attraction of cations may be pictured as in Figure 6-3.

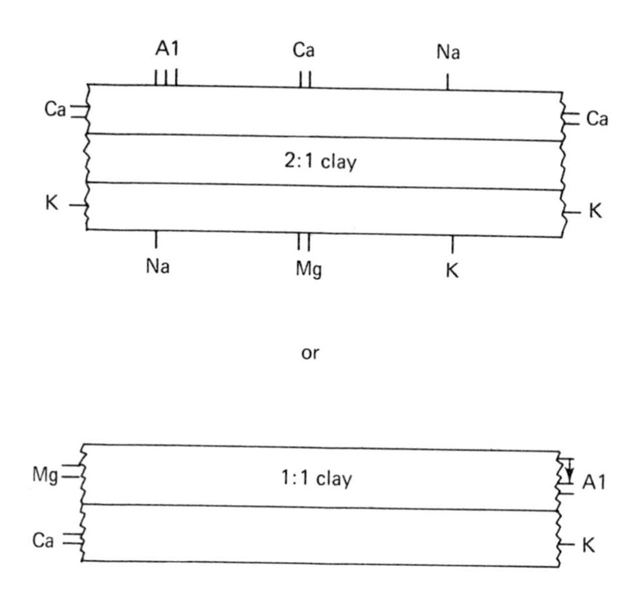

Figure 6-3. Schematic diagram of how both 1:1 and 2:1 clays might absorb and hold nutrients.

These cations are held very similar to the way a negative pole of a magnet holds the positive pole of another magnet. If we remember that in a soil system we also have water films around the soil particles, we might enlarge the previous diagram to the system depicted in Figure 6-4. In Figure 6-4 we have four soil colloids with water films around them. There are cations attached to the soil colloids and other cations in the soil solution. These cations come from fertilizers and the breakdown of soil minerals and organic matter. The aluminum has weathered from the clay and the hydrogen has come from the water.

Cation exchange takes place when one of the soil solution cations replaces one of the cations on the soil colloids. This exchange only takes place when the soil solution is not in equilibrium with the soil colloids. However, the soil solution and the colloids are seldom in equilibrium because leaching and plant uptake of cations is a continual process. Pictorially, this exchange might be presented as in Figure 6-5. As the two potassium ions (K^+) in the soil solu-

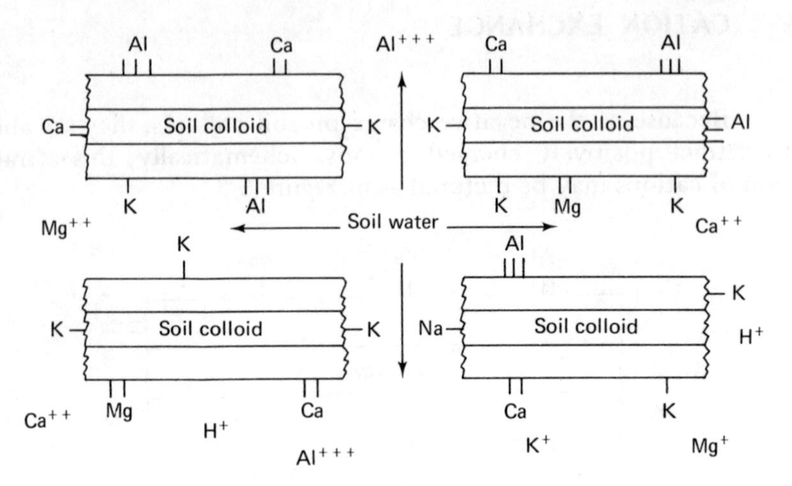

Figure 6-4. Four 2:1 colloids surrounded by a water film. These colloids control the ions in the soil water.

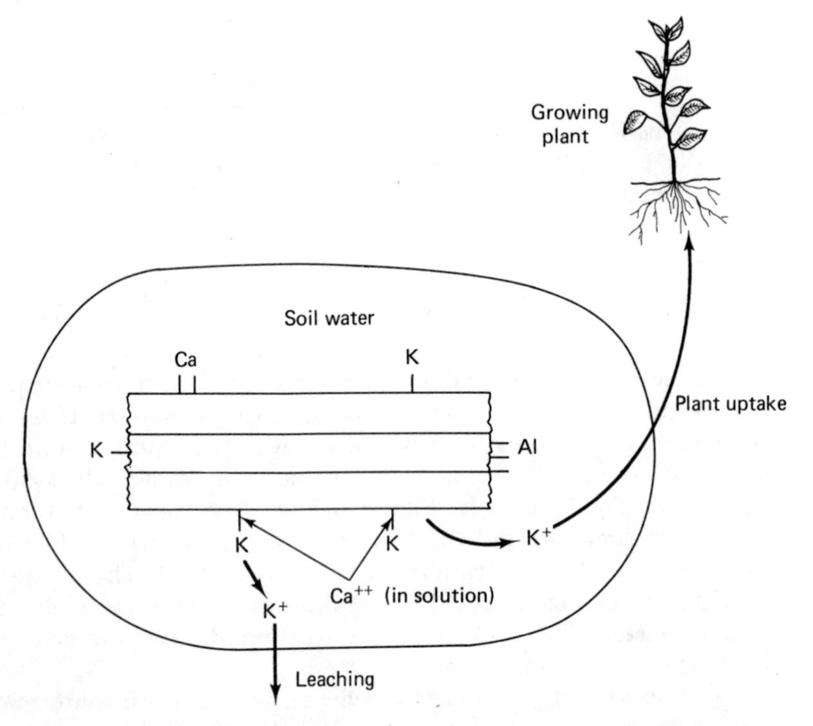

Figure 6-5. Cation exchange takes place because of leaching and/or nutrient uptake removing ions from the soil solution.

tion are taken up by the plant or leached deeper into the soil, the two K^+ ions on the soil colloid move into the soil solution. The void left on the soil colloid is satisfied by a calcium ion (Ca^{++}) from the soil solution.

CATION EXCHANGE CAPACITY

Cation exchange capacity (CEC) is simply the ability or capacity of a soil colloid to hold cations. This capacity is directly dependent on the amount of charge on the soil colloid. From our previous discussion it should be remembered that the amount of charge on the colloids is dependent on the types of colloids present.

Measurement of Cation Exchange Capacity

Cation exchange capacity (CEC) is measured in terms of milliequivalents per 100 grams of soil. A milliequivalent of any cation is that amount of cation which is required to replace one milliequivalent (or one milligram) of hydrogen. A milliequivalent weight is 1/1000 of an equivalent weight. Thus, if the atomic weight of hydrogen is 1 and the valence is 1 then the milliequivalent weight must be 1 milligram. Therefore it would take 20 milligrams of calcium to replace 1 milligram of hydrogen. The following practical example may clarify this. If a soil had 1 milliequivalent of cation exchange capacity per 100 grams of soil, then it could hold the following amounts of calcium:

$$\frac{.020 \text{ g calcium}}{100 \text{ g soil}} = \frac{X \text{ pounds of calcium}}{2{,}000{,}000 \text{ pounds of soil}} = 400 \text{ lb. of calcium/A}$$

where: .020 g calcium = the milliequivalent weight of calcium

2,000,000 pounds of soil represents the weights of 1 acre of soil to a depth of 6 inches.

Factors Affecting the Cation Exchange Capacity of Soils

There are three basic factors which affect the cation exchange capacity of a soil. These are the number of colloids present (soil texture), the type of colloids present (soil consistence), and organic matter content.

SOIL TEXTURE: In Chapter 4, soil texture was defined as the relative amounts of sand, silt, and clay in a soil. We now know that the amount of clay is really the inorganic colloidal fraction of the soil. Thus, we can see that by knowing the soil texture we can estimate cation-holding capacity.

TYPE OF CLAY OR SOIL CONSISTENCE: From the soil texture we can determine the amount of inorganic colloidal material present in the soil. However, we cannot accurately estimate cation exchange capacity without knowing the types of colloids present, i.e., relative amounts of 1:1 and 2:1 colloids. As indicated in Chapter 4, we can determine the colloidal types present by determining soil consistence.

AMOUNT OF ORGANIC MATTER: For each percent humus in the soil the cation exchange capacity is increased 2 meq/100g. Thus, we can see that in sandy soils where the cation exchange capacity is only 4 meq/100g, 1 percent organic matter present would contribute one-half of the cation exchange capacity. If we use our sense of feel to determine soil texture and consistence, and visually estimate organic matter contents from soil color we also can estimate cation exchange capacity.

Importance of Cation Exchange Capacity in Soil Management

1. It indicates the nutrient-holding capacity of a soil.
2. It determines how often and how much lime must be applied (Chapter 8).
3. Cation exchange capacity determines how crop nutrients other than lime can be applied. On very low cation exchange capacity soils, potassium may need to be side dressed during the growing season; but on high and medium cation exchange capacity soils, it can be broadcast before planting.
4. On high cation exchange capacity soils, anhydrous ammonia is the cheapest form of fertilizer N, but on low CEC soils it may leach through the soil after heavy rains or escape to the atmosphere upon application.

SUMMARY

Soil colloids may be defined as soil particles less than 0.002 mm in diameter which carry an electrical charge and are capable of staying suspended in water. Soil colloidal materials arise from soil or silicate clays, hydrous oxides, and organic residues.

The most abundant colloidal materials are the silicate clays. Basically, there are two types of clays; the 1:1 types consisting of one silica and one aluminum layer and the 2:1 types which have two silica layers and one aluminum layer. The dominant 1:1 minerals are kaolinite and halloysite and the 2:1 minerals are montmorillonite, illite, and vermiculite. Often aluminum can get in the interlayer space of vermiculite and give rise to a 2:2 clay mineral having two layers each of silica and aluminum.

Although hydrous oxides are not as abundant as silicate clays, they are of a colloidal nature and add to the nutrient-holding capacity of the soil. These materials can also carry a positive charge and are often responsible for phosphorus fixation.

Unlike silicate clays, organic colloids do not have a definite structure. They consist of organic masses of broken chemical bonds. The chemical bonds can give rise to a very high negative charge, which is capable of holding large amounts of nutrients.

Since soil colloids carry a negative charge, they can attract and exchange positively charged ions (cations). This property is called cation exchange capacity. This is a very important property because it in essence is the nutrient-holding capacity of the soil. Thus, we can say that the cation exchange or nutrient-holding capacity of a soil is regulated by the amount and type of colloidal materials present in the soil.

In addition to regulating the chemical properties of a soil, the colloidal fraction also plays an important role in the physical properties of a soil. In Table 6-1, the various physical and chemical properties of the major soil colloids are summarized. After studying this table, the student may wish to review those portions of Chapter 4 which deal with soil texture and consistence.

Soil colloids are quite important in soil management because they contribute to soil physical properties through soil consistence and soil chemical properties through cation exchange. The effect the soil colloids will have on any given soil will depend on the amount and type of colloids present. For example, montmorillonite in large quantities in a soil may cause it to be plastic, sticky, and have a very high nutrient-holding capacity. Whereas, a small amount of montmorillonite might be very helpful by increasing the water-holding capacity and reducing leaching.

For the soil manager, the only simple determinations to find the type and amounts of colloids are the soil consistence and texture measurements made by feel. By estimating the amounts and types of colloidal materials present, the manager can estimate the soil chemical and physical environment he will be facing.

Due to complex patterns of soil colloids in the United States, a detailed description of the occurrence of different types of col-

Table 6-1
Properties of Selected Soil Colloids

Type of Colloid	*Particle Size*	*Particle Surface Area*	*Stickiness*	*Plasticity*	*Ability to Shrink and Swell*	*CEC meq/100g*
Kaolinite	Large .002-.0002mm	Small	Slight	High	None	2-10
Halloysite	Medium	Medium	Slight	High	None	2-10
Montmorillonite	Very fine .0002mm	Very high	Very high	Very high	Great	80-120
Illite	Medium	High	Medium	High	Very slight	10-40
Vermiculite	Fine	High	Medium	Very high	Medium	80-100
Vermiculite-Chlorite	Fine	High	Medium	High	Medium to slight	Variable
Hydrous oxides	Fine	High	–	–	None	Low
Organic	Very fine	Very high	–	–	Very great	100-300

loids will not be attempted. Generally 2:1 types including montmorillonite and illite are present in many of the glaciated areas of the Northeast and Midwest. Kaolinite, vermiculite, and illite dominate the mountains of the Southeast. The Piedmont, adjacent to the Atlantic Coastal Plain, is dominated by kaolinite and some vermiculite. Usually kaolinite and 2:2 clay minerals dominate the Atlantic Coastal Plain. The loess soils east of the Mississippi usually contain varying amounts of illite, vermiculite, and montmorillonite.

Anyone needing assistance on the types of colloidal material present in a given area may contact the local U.S. Department of Agriculture Soil Conservation Service office and obtain assistance. Often on-site inspection is the only alternative because soils can be quite variable and small areas of 2:1 clay may appear in large areas dominated by 1:1 minerals.

REVIEW QUESTIONS

1. Define a soil colloid.
2. What is a 1:1 clay mineral? A 2:1 clay mineral?
3. What are hydrous oxides?
4. What causes the charges on organic colloids?
5. What causes cation exchange?
6. Define cation exchange capacity.
7. How is cation exchange capacity measured?
8. What factors affect cation exchange capacity?
9. Why is cation exchange capacity important in soil management?
10. How are soil colloidal properties important in soil management?
11. Review the properties of soil colloids in Table 6-1.
12. Can you define the terms in Appendix 6-1?

REFERENCES

Berger, Kermit C. *Introductory Soils*. New York: Macmillan Co., 1965, pp. 78-88.

Brady, Nyle C. *The Nature and Properties of Soils*. 8th ed. New York: Macmillan Co., 1974, pp. 71-109.

Donahue, Roy L., Shickluna, John C., and Robertson, Lynn S. *Soils: an Introduction to Soils and Plant Growth*. 3rd ed. Englewood Cliffs, N.J.: Prentice-Hall, 1971, pp. 52-67.

Foth, H. D., and Turk, L. M. *Fundamentals of Soil Science*. 5th ed. New York: John Wiley, 1972, pp. 149-78.

Appendix 6-1

Soil Chemistry Definitions

1. *Element* A single substance that cannot be subdivided into additional components.

2. *Compound* Composed of two or more elements mixed in fixed proportions.

3. *Atom* Smallest uncharged part of an element which can take place in a chemical reaction.

4. *Ion* An atom that carries an electrical charge. An ion may also be a combination of more than one atom.

5. *Cation* A positively charged ion. The cations we usually encounter in soils are Ca^{++}, K^{+}, Mg^{++}, Fe^{+++}, Al^{+++}, H^{+}.

6. *Anion* Negatively charged ions. In soils we find Cl^{-}, NO_3^{-}, $H_2PO_4^{-}$, and SO_4.

7. *Atomic Weight* The relative weights of the elements, e.g., O = 16, Ca = 40, N = 14.

8. *Valence* The number of charges on an ion.

9. *Molecular Weight* The sum of the atomic weights of

the elements found in a compound, e.g., atomic weight of sodium chloride:

Sodium	=	23
chloride	=	35
		58 = molecular weight

10. *Equivalent Weight* The weight associated with one unit of valence or the atomic weight divided by the valence. Chemicals react in equivalent weights, i.e., one equivalent weight of a substance will react completely with one equivalent of another substance.

7

Soil Acidity

"What is soil acidity?"
"Sour soils."
"Aluminum on the soil colloid."
"Hydrogen in the soil solution."

Acid or "sour" soils are one of the most important soil fertility problems in the eastern United States. The reason soils are acid is because excess H^+ ions are in the soil solution. As will be seen later, this excess of hydrogen in the soil solution is caused by aluminum entering the soil solution and liberating hydrogen ions from the soil water.

THE CHEMISTRY OF SOIL ACIDITY

In the previous chapter the concepts of colloidal charge and cation exchange were introduced. These concepts enable us to better understand the chemistry of soil acidity. As clay colloids weather and break down, the aluminum in the aluminum layer is freed and either attached to the colloidal charges (cation exchange complex) by replacing H^+ or released into the soil solution. If it is attached to the cation exchange complex, it is able to undergo cation exchange and enter the soil solution at any time. Once the aluminum enters the soil solution it reacts with water to form hydroxy aluminum compounds and free hydrogen ions. The exact aluminum compounds formed will depend on the soil acidity at the time the aluminum enters the soil solution. The reactions that aluminum can undergo are as follows:

1. $Al^{+++} + 3H_2O \longrightarrow Al(OH)_3 + 3H^+$
2. $Al^{+++} + 2H_2O \longrightarrow Al(OH)_2^+ + 2H^+$
3. $Al^{+++} + H_2O \longrightarrow Al(OH)^{++} + H^+$

If the soil solution is very acid at the time the Al^{+++} is released, it may simply stay in the solution as Al^{+++}. This is because there is a definite shortage of OH^- groups to react with it. When this happens, the Al^{+++} will compete for exchange sites and gradually reduce the cation exchange complex to nearly pure aluminum saturation. This only happens in rare situations where soils are very acid. Usually aluminum is found in the soil solution as one of the products of the previous three chemical reactions. The important points for the beginning student to remember are that soils are acid because of H^+ ions in the soil solution. These H^+ ions are the result of aluminum from the soil colloids reacting with water to give free hydrogen and hydroxy aluminum compounds, as well as certain nitrogen and sulfur reactions which will be discussed later.

TYPES OF SOIL ACIDITY

Originally soil scientists thought that soil acidity was due to hydrogen both on the soil colloids and in the soil solution. Because of this they called the H^+ in the soil solution *active acidity* and the H^+ on the soil colloids *potential acidity*. The sum of the active plus the potential acidity was called the *total acidity*. Many recent investigations have shown that there are few H^+ ions on the soil colloids. Instead, the acidity on the soil colloid is actually due to the Al^{+++} ions. Since these findings, the terms active and potential acidity have not been used greatly. However, they are still valid concepts if properly used. These concepts are best described by Figure 7-1.

The major problem with this concept is that the beginning student does not know what form of hydroxy aluminum will result. Thus, the idea of summing the active and potential acidity becomes quite complicated.

MEASUREMENT AND EXPRESSION OF SOIL ACIDITY

Acidity is measured in moles of hydrogen per liter of solution. The beginning student might think of this as the gram molecular weights of hydrogen per liter of solution. For pure chemicals ranging from highly acid to highly basic, these measurements will

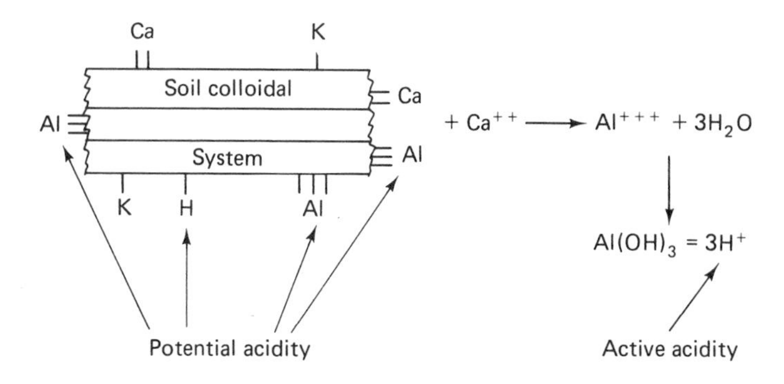

Figure 7-1. Potential and active acidity. A cation such as Ca^{++} if added to the soil system releases the aluminum from the soil colloid. The aluminum in turn reacts to form hydroxy-aluminum compounds and H^+.

range from 1 to 0.00, 000,000,000,001 of a mole of hydrogen per liter; i.e., from 1 to 10^{-14} moles of hydrogen per liter. From these figures it is quickly evident that an acidity measurement of 10^{-14} moles per liter is unwieldy and hard to express. For this reason, chemists use what is known as the pH scale to express acidity. The pH figure is nothing more than the negative logarithm of the moles of hydrogen per liter. The negative logarithm of 1 is 0 and of 10^{-14} is 14. Therefore, the pH scale ranges from 0 to 14. Zero on the scale is very acid and 14 is very basic. A pH of 7 is neutral. The pH scale is schematically depicted in Figure 7-2.

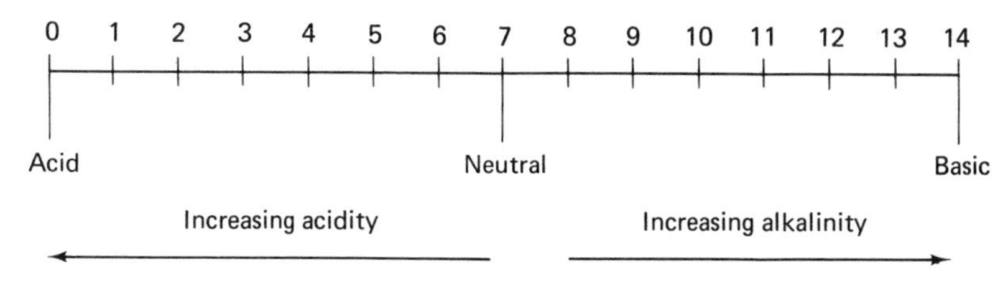

Figure 7-2. The pH scale. Common soil pH values range from about 3.8 to 8.2.

When using the pH scale, the student should remember it is a logarithm and a one-unit pH change means a tenfold change in H^+ concentration. For example, 0.001 of a mole of hydrogen per liter equals a pH of 3 while 0.0001 of a mole of hydrogen per liter equals a pH of 4.

For most soils the pH can range from around 3.5 to 8.3 depending on the cations present. In the high sodium soils of the western United States the pH may be 11 or higher.

THE PERCENT BASE SATURATION CONCEPT

Soil scientists often refer to the percent base saturation in the soil. This simply means the percent of the cation exchange complex saturated with basic cations. Basic cations are considered any cations except hydrogen and aluminum.[1] Percent base saturation really means that amount of the cation exchange capacity not holding potential acidity.

It is a very handy measure to use in expressions of soil fertility because a high percent base saturation means desirable nutrient levels and low soil acidity. Schematically the percent base saturation might be expressed as in Figure 7-3.

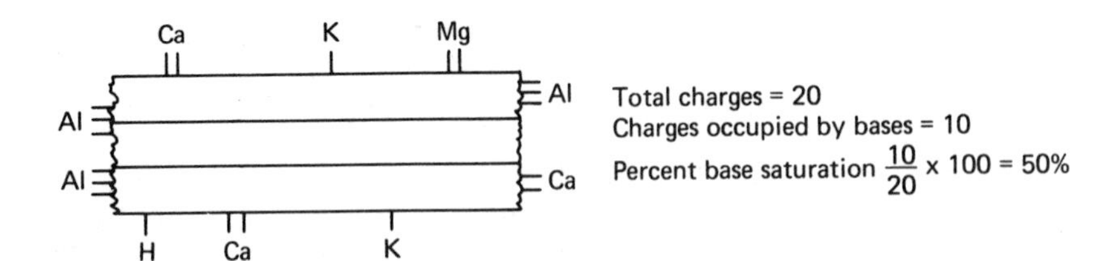

Figure 7-3. Percentage base saturation. The percentage base saturation may be envisioned as above. The actual techniques to determine PBS are quite detailed.

In the eastern United States, soils range from less than 35 percent to 90 percent base saturated. In the Southeast the base saturations will range from 35% to 50% unless heavy limestone applications have been made. Under high liming programs the Ap horizon percent base saturation may reach 80 to 90 percent. The younger soils of the North and Midwest have base saturations of greater than 35 percent and often as high as 90 percent.

SOIL ACIDITY AND SOIL MANAGEMENT

To better understand the practical implications of soil acidity, let us look at the harmful effects of acid soils, the reasons soils become acid, and the desirable pH ranges for most crops. The control of soil acidity will be discussed in the next chapter.

Harmful Effects of Soil Acidity

1. Some plants simply do not grow well at a low pH, (i.e., are not adapted).

---

[1] Actually, iron and manganese are also considered as nonbasic cations. A discussion of the effects of these cations is not within the scope of this text.

2. The activities of many of the following soil organisms are reduced:

   a. Nitrogen-fixing bacteria

   b. Bacteria which convert ammonium to nitrate

   c. Organisms which break down organic matter

3. Elements such as aluminum and manganese become so soluble they are toxic to plant growth.

4. Phosphorus and molybdenum may become insoluble and unavailable.

5. A low pH may indicate low levels of calcium and magnesium present.

Reasons Soils Become Acid

Previously we saw how hydrogen and aluminum cause soil acidity. As shown below, we see how these two cations become predominant in the soil.

Figure 7-4. The corn in the foreground is being grown on deep sand which has not been limed in several years. Note the poor growth and retarded development.

1. Leaching—Percolating water removes nutrient elements which are replaced by hydrogen and aluminum.
2. Nutrient removal by crops.
3. The use of acid-forming fertilizers.
4. The breakdown of organic matter releasing hydrogen.
5. Acids produced by growing roots.

Acidity Ranges of Various Crops

The most desirable pH ranges for many common field crops are presented in Table 7-1. Usually when the soil pH is adjusted to these ranges, the effects of other nutrients can be optimized.

Table 7-1
Desirable pH Ranges for Some Common Crops

Crop	*pH range*
Alfalfa	6.0-6.7
Azaleas and camellias	5.0-5.5*
Blueberries	4.8-5.2*
Corn	5.8-6.7
Cotton	5.8-6.4
Grasses	5.8-6.2
Irish potatoes	4.8-5.5*
Lespedeza	5.8-6.2
Peanuts	5.8-6.2
Small grains	5.8-6.2
Soybeans	5.8-6.7
Sweet and red clover	6.3-6.7
Tobacco	5.0-6.2*
General garden crops	5.8-6.2
General legume crops	5.8-6.7

*Note: May become susceptible to disease at higher pH levels.

SUMMARY

By this time the student should be aware that hydrogen is responsible for soil acidity and is the result of certain aluminum reactions in the soil. Soil hydrogen levels are expressed in terms of pH which is a negative logarithm of the hydrogen ion concentration. Because of the harmful effects of acid soils and the continual causes of soil acidity, it is important the student have a firm understanding of it.

REVIEW QUESTIONS

1. What are the aluminum reactions in soils?
2. How is soil acidity expressed?
3. What is meant by percent base saturation?
4. What are the causes of soil acidity?
5. What are the harmful effects of acid soils?
6. What are the most desirable pH ranges for common field crops?
7. Discuss the relationships between pH, active acidity, potential acidity, total acidity, cation exchange capacity, and percent base saturation.

REFERENCES

Berger, Kermit C. *Introductory Soils.* New York: Macmillan Co., 1965, pp. 137-45.

Brady, Nyle C. *The Nature and Properties of Soils.* 8th ed. New York: Macmillan Co., 1974, pp. 372-402.

Donahue, Roy L., Shickluna, John C., and Robertson, Lynn S. *Soils: an Introduction to Soils and Plant Growth.* 3rd ed. Englewood Cliffs, N.J.: Prentice-Hall, 1971, pp. 277-81.

Foth, H.D., and Turk, L.M. *Fundamentals of Soil Science.* 5th ed. New York: John Wiley, 1972, pp. 179-88.

McVickar, Malcolm H. *Using Commercial Fertilizers: Commercial Fertilizers and Crop Production.* 3rd ed. Danville, Ill.: Interstate Printers and Publishers, 1970, pp. 145-52.

8

Liming

"How do you control soil acidity?"

"Lime your soils."

"Remove the hydrogen and aluminum from the soil system."

"Raise the soil's pH."

A *lime* is defined as any calcium-containing material which can be added to the soil to increase the pH. This pH increase takes place because some of the hydrogen ions in the soil solution are removed. To remove these hydrogen ions we must reverse the processes which make a soil acid. (See chapter 7.) To do this it is necessary to remove the hydrogen (active acidity) from the soil solution and neutralize the aluminum (potential acidity) on the soil colloids. Schematically, this reaction might be presented as in Figure 8-1.

From Figure 8-1 it can be seen that the goals of liming are to achieve a neutral soil colloid, inactivate the aluminum, and remove the hydrogen as water. It is the goal of this chapter to outline how these reactions take place.

THE REACTIONS OF COMMON LIMING MATERIALS

There are many liming materials available to the soil manager and landscaper. The most common of these are: calcitic limestone, dolomitic limestone, burned lime, and hydroxide of lime. Other materials such as marl, basic slag, wood ashes, and industrial

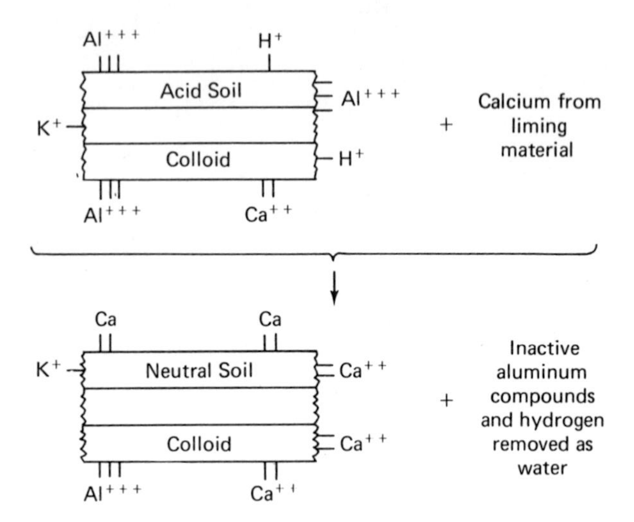

Figure 8-1. Generalized limestone reaction.

sludges are used as limes to a much lesser extent. The reactions of these materials are as follows:

Calcitic Limestone

Of all the liming materials available, the most common one used is probably calcitic limestone. It is ordinarily referred to simply as limestone. The chemical composition of it is calcium carbonate ($CaCO_3$) and it is often quite pure even though it is a naturally occurring mineral. Because it is quite common, the reactions it undergoes in the soil will be used as a reference for the other materials. The chemical reactions of this limestone are somewhat detailed and should be studied with care. These reactions are outlined in Figure 8-2. The reactions in Figure 8-2 take place because the Ca^{++} and $CO_3^{=}$ ions are added in large quantities. If they were added in small amounts, there would be little effect on the soil pH. The amounts that will need to be added will depend directly on the soil's cation exchange capacity. If the soil is very sandy and has a low CEC, the amount of limestone needed to change the pH may not be more than 1,000 to 1,500 pounds per acre. On high 2:1 clays or organic soils it may take 10 to 20 tons of limestone per acre to initiate a pH change of any magnitude.[1] The general pH change from liming with different textural classes is shown in Figure 8-3.

---

[1] The resistance of a soil to a pH change when limestone is added is called the *buffer capacity*. The student should be aware that this buffer capacity is directly related to the CEC.

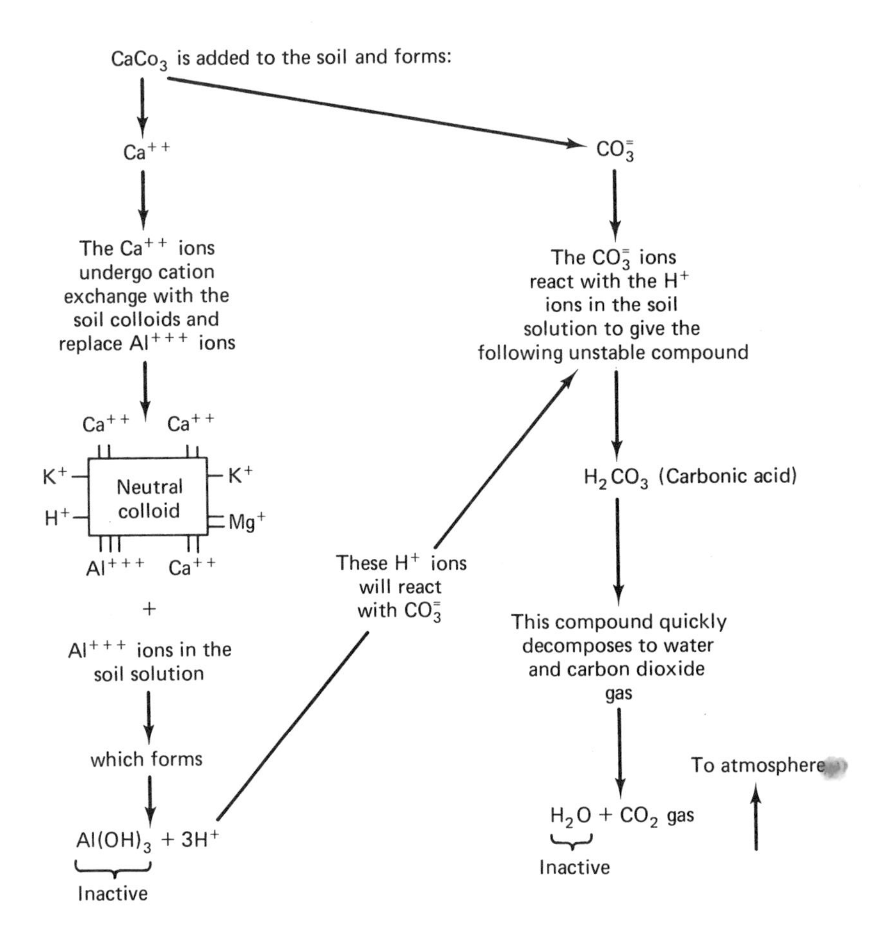

Figure 8-2. The reactions of calcitic limestone in the soil. The overall net effects of this reaction are neutral soil colloids and a reduction in the relative number of H^+ ions in the soil solution. As a consequence of this, the pH of the soil is raised.

Dolomitic Limestone

Although calcitic limestone is probably the most common limestone used across the nation, dolomitic limestone is the most common in the tobacco-producing states. This is because dolomite is a calcium magnesium carbonate rather than a pure calcium carbonate. When used in tobacco rotations, it furnishes needed magnesium to the tobacco. The reactions of dolomitic limestone in the soil are the same as calcitic, except both Ca^{++} and Mg^{++} ions are responsible for exchanging with the Al^{+++} ions held on the acid soil colloids. Because the crystalline structure of dolomite is

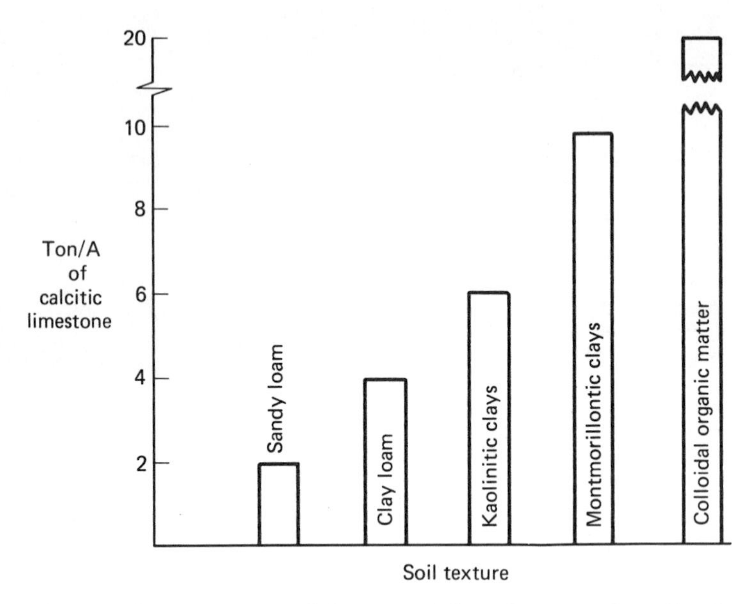

Figure 8-3. Amounts of lime needed to increase the pH of various soils from 4 to 6. This graph is a generalization from the results obtained in several beginning soils laboratories at North Carolina State University, Raleigh, North Carolina.

different than calcite, it breaks down (dissolves) more slowly in the soil. This allows dolomite to last longer in sandy, low CEC soils. It also increases the length of time it takes to raise the pH. This should be considered if an immediate pH increase is needed.

Burned Lime

Burned lime is a calcium oxide (CaO) compound. When it enters the soil solution, it reacts with water to form $Ca(OH)_2$, or calcium hydroxide. It then dissolves or dissociates to form Ca^{++} and OH^- ions. The Ca^{++} ions replace the Al^{+++} ions on the soil colloids as in the calcitic reactions in Figure 8-2. The OH^- ions react directly with H^+ ions in the soil solution and water is formed to remove the excess H^+ ions. In general, the reactions of calcium oxide with acid soil colloids are the same as the calcitic reactions; except that the H_2CO_3 reaction is absent and water is formed directly from the $H^+ + OH^-$ reaction.

Burned lime reacts quite quickly in the soil and has the ability to neutralize quite a lot of hydrogen due to its low atomic weight per molecule. Because of its chemical attraction for water molecules, burned lime is caustic (burns the skin) and unpleasant to handle. It also tends to draw moisture which makes it rather hard to store for long periods.

The process for producing burned lime consists of heating a magnesium or calcium carbonate compound until carbon dioxide is driven off. This leaves behind the oxide form. Common names for burned lime include burned lime, quick lime, burned oyster shell lime, and simply oxide in some areas. Because it requires the heating process, burned lime is usually quite expensive compared to calcite and dolomite. The major value of burned lime is its rapid speed of reaction and high neutralizing value.

Hydroxide of Lime

Hydrated lime is simply burned lime with water added to it to form calcium and/or magnesium hydroxide compounds, $Ca(OH)_2$ and/or $Mg(OH)_2$. The reactions of these limes in the soil are the same as those for burned lime except the initial reaction to form $Ca(OH)_2$ has already taken place. They are also quite caustic or unpleasant to handle. Their major importance lies in their reaction speed. Common names for hydroxide of lime are hydrated lime and slaked lime.

Miscellaneous Liming Materials

There are numerous miscellaneous liming materials such as marl, ground sea shells, basic slag, wood ashes, sludges, silicate slag, etc. These materials may be either carbonates, oxides, or hydroxides. Their reactions in the soil will be the same as those for the materials previously mentioned. Be careful when using them because they may have very low neutralizing values. If the neutralizing values are low, they may become quite expensive compared to calcite or dolomite.

FACTORS THAT AFFECT THE ACTION OF LIME IN SOILS

With the previous discussion in mind, we are now ready to look at four factors which affect the action of any lime in a soil. These factors are important to the soil manager because they are items he can control through management decisions.

Particle Size

Because they have more surface area per unit volume or unit of weight, fine limestone particles will dissolve faster in the soil and raise the pH faster. However, they will not last as long or give a stable pH over a period of time. The proper limestone should probably contain both some coarse and fine particles. To measure the fineness of limestone particles, they are passed through a screen. The number of openings per square inch is known as the *mesh* of the screen. The laws regulating the fineness of grind of limestone vary with each state. However, they are usually defined as a very large percentage passing a ten-mesh screen and a smaller percentage passing a fifty-to-one-hundred-mesh screen.

Degree of Mixing

For a limestone to react and be effective in the soil, it should be well mixed so that the soil and lime particles are in contact. The only proper way to do this is to apply the lime prior to tillage. On small, very acid areas it might even be wise to plow down part of the limestone and place the rest on the surface after plowing and mix by disking. This type of mixing would probably not be feasible on large areas.

In recent years, many investigators working in the area of no-tillage crop production have recommended limestone be applied to the soil surface and allowed to leach downward. The authors take a very dim view of this practice. Our conclusions are based on research conducted with surface applied limestone on clay soils. From this work it was found that surface applied limestone moved downward less than 1 cm in eighteen months. This slow movement is insufficient to keep the soil root zone at a satisfactory pH level.

It is necessary to apply limestone to the surfaces of lawns and turfs which are in good condition and well established. It is not wise, however, to try to establish a turf on an acid soil without mixing the lime with the soil. Surface application of lime should also be made on established pastures. When surface applying limestone to grasses and turfs, it is wise to make smaller, more frequent applications, rather than larger applications less often.

Relative Neutralizing Value of the Limestone

When deciding on types of limestone to purchase, it is wise to consider the neutralizing value of the lime. The neutralizing value of a lime is the amount of soil acidity it will neutralize as compared to pure calcium carbonate. The neutralizing value is often referred to as the calcium carbonate equivalent. For ease of understanding, pure $CaCO_3$ is assigned a value of 100 and all other limestones are ranked accordingly. The neutralizing values of selected limestones are given in Table 8-1.

Table 8-1
Neutralizing Values for Selected Liming Materials

Liming Material	*Neutralizing Value**
Calcium carbonate	100
Dolomitic limestone	100
Calcitic limestone	95
Hydrated lime	125
Burned lime	160
Marl	70
Basic slag	60
Wood ashes	45

*For all materials other than pure calcium carbonate, the neutralizing values vary from 10 to 15 percent.

Magnesium Content

When purchasing lime, always consider the magnesium content. On magnesium-deficient soils, dolomitic limestone or some liming material with magnesium added will be the quickest, easiest way to increase the magnesium levels. This is quite important on tobacco soils and soils with very low magnesium levels as compared to calcium levels.

FACTORS TO CONSIDER WHEN DETERMINING AMOUNT OF LIME TO APPLY

For the soil manager to make a sound decision on the amounts of lime to apply, he needs a soil test. In the absence of a soil test he can, however, make a relatively reliable guess. When he makes this guess, he should consider the following:

1. Present soil pH if known.
2. Cation exchange capacity of the soil (usually determined by texture and consistence).
3. Crop to be grown.

Some general guidelines are as follows:

1. On sandy soils 1,000 to 1,500 pounds of dolomitic limestone should be added every two years.

2. On clay loam soils, 4,000 to 5,000 pounds every three to four years may be necessary.

3. On 2:1 clay soils 6,000 to 8,000 pounds may be added every four to six years.

4. On organic soil, have the soil tested (it is unwise to guess.

These guidelines are not a replacement for a soil test. They are only given to guide the manager.

LIMESTONE APPLICATIONS

The most common method of limestone application is a double spinner spreader mounted on a truck or trailer chassis (See Figure 8-4). These spreaders are quite rapid and fairly accurate, provided they are kept clean and in good repair.

A few years ago, much limestone was bagged and sold dry. If it was sold in bulk it was usually kept dry. In recent years it has become common to stockpile limestone along the edges of fields or any area convenient to the areas to be limed. The lime in these stockpiles is exposed to the weather and often called wet lime. There is no difference in the wet or dry liming materials. However, the wet lime contains some water and should be applied such that the weight of the water is considered in application rates. Wet lime also tends to cake in the spreader box and may need to be watched carefully to insure uniform application rates. Dry lime can be very dusty and cause a great deal of air pollution during application. The wet lime gives very few dust problems. In any case both materials can be used quite satisfactorily.

SUMMARY

A lime is defined as any calcium containing material which can be added to the soil to increase the pH. The pH increase is due to the components of the lime replacing aluminum-and/or hydrogen-forming insoluble aluminum compounds and water.

Figure 8-4. A very common limestone applicator is the double spinner spreader.

Common liming materials are calcitic and dolomitic limestone, burned lime, hydroxide of lime, and many miscellaneous materials. The most suitable material depends on the magnesium needs of the soil, the reaction speed desired, and the cost of each of the materials based on their relative neutralizing values. The factors which affect the action of any lime in soils are particle size, degree of mixing, neutralizing value of the limestone, and the magnesium content.

When considering the amounts of lime to apply, consider the crop to be grown, the present pH, and the cation exchange capacity. Different crops require different acidity levels. The amount of lime needed to raise the pH will vary from sands to clays and among different types of clays.

Either wet or dry lime can be easily applied with a double spinner spreader. Wet limes have a few application problems but cause less dust pollution than dry lime.

REVIEW QUESTIONS

1. How does a lime correct soil acidity?

2. What reactions do the various liming materials (oxides, hydroxides, and carbonates) undergo in the soil?

3. What factors affect the actions of lime in the soil?

4. What factors should be considered when determining the amount of lime to apply?

5. Define relative neutralizing value.

6. What is the basis for most state lime laws relative to fineness of grind?

7. What problems are associated with wet and dry lime applications?

REFERENCES

Berger, Kermit C. *Introductory Soils*. New York: Macmillan Co., 1965, pp. 146-66.

Brady, Nyle C. *The Nature and Properties of Soils*. 8th ed. New York: Macmillan Co., 1974, pp. 404-21.

Donahue, Roy L., Shickluna, John C. and Robertson, Lynn S. *Soils: an Introduction to Soils and Plant Growth*. 3rd ed. Englewood Cliffs, N.J.: Prentice-Hall, 1971, pp. 281-96.

Foth, H.D., and Turk, L.M. *Fundamentals of Soil Science*. 5th ed. New York: John Wiley, 1972, pp. 189-202.

9

Organic Matter and Soil Organisms

"What are the functions of soil organisms?"

"Some are harmful—but many are beneficial."
"The breakdown of organic matter and nutrient release."
"Organic gardening?"

The subjects of soil organic matter and soil organisms are quite large and complicated. The goal of this chapter is to simply give the student a general overview of these subjects.[1]

ORGANIC MATTER

In this text the specific properties of organic matter are not treated separately. The formation of organic soils is covered in Chapter 3, and the chemical and physical properties are covered in Chapters 4, 5, and 6, along with the mineral soils. This chapter is devoted to the definition, composition, and importance of organic matter.

Definition of Organic Matter and Humus

Soil organic matter may be defined as any living or dead plant or animal materials in the soil. Organic matter is often confused with humus. *Humus* is the substance left after soil organisms have become modified from the original organic matter to a rather

[1] Nyle C. Brady, *The Nature and Properties of Soils*, 8th ed. (New York: Macmillan Co., 1974), pp. 111-161.

stable group of decay products. In other words, humus is the colloidal remains of organic matter.

Composition of Organic Matter

Because organic matter consists of either living or dead plants and animals, it must contain all the nutrients needed for the growth of these organisms. The amounts and types of these nutrients are directly dependent on the original source.

In raw (undecomposed) organic matter there are four major groups of organic compounds contributed to the soil. These are carbohydrates; lignins; proteins; and fats, waxes, and resins. Of these four compounds, the carbohydrates and proteins are probably the most important and the most readily decomposed. These two constituents are also the greatest contributors of soil nutrients, such as nitrogen, sulfur, and phosphorus. Lignin is a very resistant compound which persists in the soil as one of the main components of humus. Fats, waxes, and resins are resistant compounds which contribute sulfur and phosphorus to the soil. Along with these four major compounds are a host of minor compounds which are beyond the scope of this text.

As the four major compounds previously mentioned decompose, several nutrients are released into the soil. Some of these nutrients are:

1. Carbon and carbon compounds such as carbon dioxide (CO_2) and carbonates ($CO_3^{=}$)
2. Nitrogen—many forms
3. Sulfur—many forms
4. Phosphorus
5. Hydrogen and hydroxide
6. Mineral elements such as potassium, magnesium, and calcium
7. Microelements such as zinc, manganese, etc.

Unless the organic matter is decomposed, these nutrients will remain in the soil but in unavailable forms. The decomposition of organic matter is carried out by soil organisms.

SOIL ORGANISMS

Organisms in the soil can be divided into two major groups. These are the animal organisms and the plant organisms. In both groups we find many different organisms varying in size from rather large to microscopic. The contributions of these organisms to the soil vary from very beneficial to very harmful. Earthworms move through the soil providing aeration and excellent soil physical characteristics. Mice and moles, however, can move through the O and A horizons and cause havoc to lawns and flower gardens. Likewise, several microscopic organisms contribute greatly to the soil while several other organisms cause diseases, poor plant rooting, etc. Depending on the soil conditions present, organisms can range in numbers from a few to billions. Because of the scope of this text, no attempt will be made to present a classification of these organisms. Only the factors which affect their presence in the soil and some of their more important reactions will be noted.

Factors Affecting the Kinds and Numbers of Organisms in the Soil

For any type of organism to grow and multiply in the soil, the following environmental conditions must exist. If these conditions are not present, the numbers of organisms will decrease and many beneficial reactions will slow down or cease.

1. Adequate aeration—If soil aeration is not sufficient, many organisms will die or become dormant.

2. Sufficient moisture content—Soil moisture must be adequate to meet the needs of the organism. If the soil is too dry, the organisms will fail to grow and reproduce. If the soil is too wet, aeration will be reduced.

3. Adequate temperature ranges—If the soil becomes too hot, the organisms may die. If it is too cold, they will become dormant.

4. Adequate food source—For an organism to thrive and multiply, it must have the proper food source. This food source is organic matter of some type. The types of organic matter present will determine the types of organisms present.

5. Adequate soil pH—Certain organisms grow and operate only at certain soil acidity levels. If the soil becomes too acid, many beneficial organisms will stop operating.

6. Proper competing organisms present—Most organisms are quite particular as to the other types of organisms around them. Certain organisms can be controlled by simply planting a crop which they cannot tolerate.

As previously stated, many organisms carry out several beneficial reactions in the soil. To take advantage of these reactions, the six items previously listed must be adjusted properly.

Reactions Carried Out by Soil Organisms

Of course, one of the prime functions of soil organisms is the decay or oxidation of organic matter by simply removing it from the surface of the earth. If general organic matter decay didn't take place, we would have long ago been buried under tons of waste materials. However, during the decay process, there are several highly beneficial reactions taking place. The following group of reactions are quite important in soils:

1. Nonsymbiotic nitrogen fixation—Organisms living in the soil independent of other organisms fix 10-20 pounds of nitrogen per acre per year from the atmosphere and contribute it to the soil.

2. Symbiotic nitrogen fixation—Organisms living in the nodules of legumes fix nitrogen from the atmosphere and contribute it to the soil when the plant dies. This fixation can be as high as 300 pounds of nitrogen per acre per year.

3. Ammonification—In the breakdown of organic matter, certain organisms free ammonium (NH_4^+) to the soil. This is quite an important reaction since it is the mechanism whereby all organic nitrogen is released to the soil.

4. Nitrification—Organisms convert ammonium (NH_4^+) to nitrate (NO_3^-). This reaction is also very important because excess ammonium in the soil can be quite harmful to plants such as tobacco and watermelons. This reaction takes place rapidly in warm weather and is capable of converting large amounts of ammonium. As this conversion takes place, some of the H^+ ions from the NH_4^+ are left in the soil solution and reduce the soil pH. Because this conversion takes place so rapidly in warm weather, we should watch soil acidity levels very closely when applying large amounts of organic matter or ammonium fertilizers. The nitrification reaction is:

$$NH_4^+ + 2O_2 \xrightarrow{\text{Organisms}} NO_3^- + H_2O + 2H^+$$

or:

$$\text{ammonium} + \text{oxygen} \xrightarrow{\text{Organisms}} \text{nitrate} + \text{water} + \text{hydrogen}$$

5. Phosphorus mineralization–Selected soil organisms convert organic phosphorus to orthophosphates. The uptake and conversions of phosphorus will be discussed more fully in the next two chapters.

6. Sulfur conversions–Most soil sulfur is held in complex organic forms. Organisms take this sulfur and convert it to sulfate ($SO_4^=$), which can be used by plants.

7. Other reactions–During the decay processes, the organisms release many other elements to the soil, such as calcium, magnesium, potassium, and micronutrients.

Other Processes Carried on by Organisms

A point that is commonly overlooked when discussing organisms is their food source during organic matter decay. For organisms to live and decay organic matter, they must have a food source. If this food source is not available in the product being decayed, these nutrients will be taken from the soil during the decay process. When this happens, the organisms are competing with higher plants for nutrients. This can cause nutrient-deficient higher plants until the organisms die and relinquish their nutrients to the soil. Because of this, the soil manager or land user should add extra fertilizer nutrients if he wishes to decay organic matter during a cropping season.

Another beneficial effect of soil organisms is their ability to promote soil aggregation. This effect was covered in Chapter 4.

THE CARBON-NITROGEN RATIO

The carbon-nitrogen ratio is a term often misunderstood by beginning students. It simply refers to the ratio of carbon to nitrogen in soils and organic matter. For most stable soil systems, this ratio is around 12:1 (twelve parts carbon to one part nitrogen). The importance of this ratio lies in the fact that it is a measure of the speed with which organic materials will decay. Many soil organisms need quite a lot of nitrogen to break down organic matter. If a wide ratio organic matter, say 90:1 (sawdust, tree bark, etc.), is added to the soil, then the organisms must take much nitrogen from the soil to be able to bring this ratio down to 12:1. This means that unless additional nitrogen is added with wide C:N ratio organic materials, they either will not decay or will cause severe nitrogen deficiencies. This is why it is often wise to plow additional nitrogen down with a heavy crop of corn stalks.

ORGANIC MATTER MAINTENANCE

From the previous discussions it should be obvious that we are continually adding organic matter to the soil and it is being broken down by soil organisms. Because this process tends to simply go in a circle, many persons often wonder if it is really worthwhile to try and add organic matter to the soil. The answer to this question is that we should try and maintain natural soil organic matter levels. If we simply forget about organic matter, we will lose all of the beneficial effects it has on soil chemical and physical properties. If we try to greatly increase organic matter levels, we will find that some of the soil organism populations will greatly increase and it will be destroyed anyway. Thus, we simply recommend organic matter maintenance. This can be done as follows:

1. Keep grasses in the crop rotation.
2. Return all crop residues to your fields.
3. Cultivate no more than necessary as it will greatly increase organic matter destruction.
4. Control erosion to avoid soil and organic colloids from washing or blowing away.
5. Use cover crops whenever possible.
6. Return all manures to the soil.

By following these practices the manager can maintain the organic content of his soils quite well. If he tries to increase them any more than this, it may become very expensive. These expenses will come both from the cost of the practices and from the revenue lost by not growing high-value row crops.

ORGANIC GARDENING

In the past few years there has been a very serious movement toward nonchemical food production by both novice gardeners and serious food producers. To many growers the organic gardening techniques are closer to religion than science. To other growers, these techniques are based on a complete understanding of organic matter and organisms and their role in the soil. High fertilizer costs nearly dictate that any food producer take advantage of the nutrients held in organic matter whenever economically possible.

If you try organic gardening, there are a few simple rules you should observe before you can expect success. The first thing to do is expel the myth that there are such things as "organic nutrients." Plants only take up nutrients in certain forms irrespective of their source. Thus, one pound of phosphorus from organic matter is the same as one pound of phosphorus from superphosphate provided organic decay has released the phosphorus from the organic matter.

The second item to remember is that leaching and nutrient removal will be greater than the residue from a crop. Thus supplemental organic matter with a narrow carbon-nitrogen ratio will need to be added. If supplemental organic matter or mineral nutrients are not supplied to cover leaching and prior crop removal, poor results can be expected.

After exploding the organic nutrient myth, and determining from soil tests the nutrients needed for a crop, you are ready to apply organic matter to your garden. If the factors that affect the decay organism in the soil are adjusted to optimum conditions, you can expect these organisms to react with the organic matter. The net result should be nutrient release from the organic matter and excellent plant growth. As a side benefit you may also enhance your soil structure and tilth.

SUMMARY

Soil organic matter may be defined as any living or dead soil organism. Humus is the colloidial remains of organic matter. In undecomposed organic matter we find the following four major organic compounds: carbohydrates; lignin; proteins; and fats, waxes and resins. In these compounds we find all the nutrients essential for plant growth. Provided we adjust the soil environment for optimum soil organism development, we can release these nutrients for plant use.

Several processes are carried out by soil organisms. If these processes are understood, we can maintain the proper soil organic matter levels for plant growth. A detailed understanding is quite essential for total organic food production.

REVIEW QUESTIONS

1. What is organic matter? Humus?

2. What are the four major components of organic matter? What do they contribute to the soil?

3. What nutrients are released to the soil as organic matter decays?

4. What factors affect the kinds and numbers of organisms in the soil?

5. What important reactions are carried out by soil organisms? Do you understand these reactions?

6. What is the carbon nitrogen ratio? What does it tell us?

7. Should we try to increase, decrease, or maintain organic matter levels? Why?

8. Is organic crop growth possible? How?

REFERENCES

Berger, Kermit C. *Introductory Soils*. New York: Macmillan Co., 1965, pp. 35-55.

Brady, Nyle C. *The Nature and Properties of Soils*. 8th ed. New York: Macmillan Co., 1974, pp. 111-43, 534-51.

Donahue, Roy L., Shickluna, John C., and Robertson, Lynn S. *Soils: an Introduction to Soils and Plant Growth*. 3rd ed. Englewood Cliffs, N.J.: Prentice-Hall, 1971, pp. 176-221, 414-25.

Foth, H.D., and Turk, L.M. *Fundamentals of Soil Science*. 5th ed. New York: John Wiley, 1972, pp. 97-139.

McVickar, Malcolm H., *Using Commercial Fertilizers: Commercial Fertilizers and Crop Production.* 3rd ed. Danville, Ill.: Interstate Printers and Publishers, 1970, pp. 271-76.

10

Plant Growth and Nutrition

"What should I feed my crops?"

"Macronutrients from the air and fertilizer—C, H, O, N, P, K, Mg, Ca, S."

"Micronutrients—the trace elements."

One of the most important challenges faced by a person responsible for soil management is to create a soil environment for optimum plant development, whether this be for salable products, feed stuffs, or for protection of the environment. To create this environment, it is necessary to have a reasonably good understanding of plant growth processes. This understanding of plant growth and a knowledge of the materials in the previous chapters of this text should help you manipulate the crop-soil system to achieve a high degree of efficiency. As one examines growth processes, it may be possible to alter and improve plant growth for the benefit of mankind.

PROCESSES DURING PLANT GROWTH

One of the most important processes in nature is photosynthesis. Carbon dioxide and water in the chlorophyll-bearing cells of plants react with sunlight (radiant energy) to manufacture sugar. Neither man nor animal can secure food except through this process of sugar synthesis and other processes of food manufacture which are dependent upon this supply of sugar. Man is also dependent on this process for his fuel supply. Coal, petroleum, and natural gas were made from plants that lived in the past.

After sugar is formed it may be utilized at once by the plant to produce energy (from the process called respiration) and to build new cells and tissues (from the process called assimilation), or it may be transformed into starches, fats, and proteins and stored in the roots, stems, leaves, fruits, and seeds. Man has long recognized these latter activities of plants, and he continually makes use of those parts of the plants in which these foods have been stored.

To increase plant growth we must increase the photosynthetic process, i.e., make it a more efficient process. This is done by increasing available soil moisture (reducing drought stresses to a practical minimum), increasing the absorption of radiant energy (sunlight) and carbon dioxide, and supplying enough and correct proportions of the proper nutrients. Increasing available soil moisture was covered in Chapter 5. The efficient use of radiant energy is usually modified by the plant breeder and the efforts of agronomists and crop producers in their arrangement of plants through optimum spacings and populations within fields. Carbon dioxide supplies, to achieve greater photosynthetic build up, are increased by organic matter additions, proper soil aeration, supplementally supplying carbon dioxide (limited use in greenhouses, orchards, and gardens), plant breeding, and plant spacings. Soil nutrient levels and immediately available nutrient supplies are controlled by the soil manager. To understand what is within his control, it is necessary to know the elements essential for plant growth.

ELEMENTS (NUTRIENTS) ESSENTIAL FOR PLANT GROWTH

There are sixteen elements, and possibly a seventeenth, which are essential for plant growth. For convenience of learning, the elements may be divided into two groups. One group, called macronutrients, is used in relatively large quantities. A second group, although utilized in a much lesser amount, is just as important and is called micronutrients. Both groups of nutrients, along with some miscellaneous elements, are listed in Table 10-1.

Sometimes the macronutrients are divided into two groups called primary and secondary nutrients. Nitrogen (N), phosphorus (P), and potassium (K) are called primary elements, and calcium (Ca), magnesium (Mg), and sulfur (S) are called secondary elements.

In this text it is not possible to describe the deficiency symp-

Table 10-1
Elements Essential to or Related to Plant Growth

Macronutrients	*Element Symbols*	*Comments*
1. Carbon	C	Supplied by air and water.
2. Hydrogen	H	
3. Oxygen	O	
4. Nitrogen	N	Supplied by atmosphere and fertilizers.
5. Phosphorus	P	Supplied by soil minerals and fertilizers.
6. Potassium	K	
7. Magnesium	Mg	
8. Calcium	Ca	
9. Sulfur	S	
Micronutrients		
1. Iron	Fe	
2. Manganese	Mn	Supplied by soil minerals and fertilizers; can cause crop failures if absent; can become toxic to crops if present in large quantities.
3. Boron	B	
4. Molybdenum	Mo	
5. Copper	Cu	
6. Zinc	Zn	
7. Chlorine	Cl	
Miscellaneous Elements		
1. Sodium	Na	May substitute for potassium.
2. Silica	Si	May accumulate in some plants.
3. Cobalt	Co	Seldom needed except by some legumes.
4. Aluminum	Al	Have never been shown essential for plant growth and can be harmful if present in large quantities.
5. Iodine	I	

toms of each of these elements, their functions in plants, and tests for these elements. Several of the references listed at the end of this chapter give this information and include color photographs.

Visual symptoms of a deficiency are helpful in determining specific nutrient needs, but often a visual symptom occurs only after the plant is critically deficient and stunted.

More useful information about nutrient needs of plants can be determined by soil tests and supplemental information from plant analyses. Growers who properly collect soil samples from fields to be planted to a particular crop and submit these samples to a soil testing laboratory will receive nutrient application suggestions. These nutrient suggestions are to be considered as an amount beyond that which the soil can provide to achieve high yield of plant parts (roots, stalk, leaves, and grain). Chemical analyses of selected plant parts (usually recent, fully developed leaves) will indicate whether adequate nutrients are present within the plant. Plant analyses are very helpful for determining micronutrient needs since soil determinations are frequently complicated and do not always give very reliable information.

Nearly all states in the United States have one or more tax-supported soil testing laboratories. There are also many privately owned laboratories and some owned by fertilizer manufacturing companies. Laboratories for plant analyses are not as common; however, more recognition is being given to plant analyses as an additional guideline in achieving high crop yields.

PLANT COMPOSITION

The elemental content of many economic crop plants has been widely studied. Table 10-2 gives typical amounts of nutrients contained in crops and also an average amount removed when the harvested portion is carried from the field.

Plant removal of fertilizer-supplied nutrients should be considered in planning and assessing the adequacy of a fertilizer program. Records of nutrients added in fertilizers, along with those removed by harvest of the crop, must be supplemented by leaching losses, fixation by the soil, erosion of soil, and the nutrients becoming available from the soil. These records should be used as a guide to balance crop removal and fertilizer applications. Furthermore, plants frequently take up more of a nutrient than needed for growth. Luxury consumption results in a much higher level of the nutrient in the plant than is necessary for optimum plant growth. Most crops require nutrients in the following proportions:

$$N \geqslant K > Ca > Mg \geqslant P \geqslant S$$

(Note: $>$ means greater than and $\geqslant$ means greater than or equal to)

Table 10-2
Approximate Plant Nutrient Content (lb/acre)
and Removal by Harvest

Crop	*Yield per Acre*	*Nitrogen* (N)	*Phosphate* (P_2O_5)	*Potash* (K_2O)
Corn	100 bu. grain	90	35	27
	4,500 lb. stover	45	20	110
		135	55	137
Cotton	750 lb. lint and			
	1,150 lb. seed	45	18	24
	1,500 lb. stalks	30	10	30
		75	28	54
Peanuts	3,000 lb. nuts	110	15	20
	4,800 lb. vines	110	30	100
		220	45	120
Soybeans	40 bu. beans	128	32	56
	1,760 lb. straw	20	8	40
		148	40	96
Tobacco	2,800 lb. leaves	60	15	90
	Stalks	35	10	100
		95	25	190
Alfalfa	4 tons	180	40	180
Coastal Bermuda	8 tons	185	70	270
Red clover	2.5 tons	100	25	100

Reprinted in part by permission from *The Fertilizer Handbook*, 1972, a publication of the Fertilizer Institute, Washington, D.C.

Figures will vary with season, plant variety, and soil fertility level.

FORMS OF NUTRIENTS THAT MAY BE TAKEN UP BY PLANTS

Plants never take up simple elemental forms of nutrients. The nutrient must be in some ionic form. The usual ionic forms are presented in Table 10-3. By being aware of the form of the essential elements, it is possible to tell how they will be held by the soil and how susceptible they are to leaching.

Table 10-3
Ionic Forms of Plant Nutrients

Nutrient	*Symbol*	*Cation*	*Anion*
Nitrogen	N	NH_4^+	NO_3^-
Phosphorus*	P		$H_2PO_4^-$
Potassium	K	K^+	
Calcium	Ca	Ca^{++}	
Magnesium	Mg	Mg^{++}	
Sulfur*	S		SO_4^{-}
Iron	Fe	Fe^{++}	
Manganese	Mn	Mn^{++}	
Boron*	B		$H_2BO_3^-$
Copper	Cu	Cu^{++}	
Zinc	Zn	Zn^{++}	
Chlorine	Cl		Cl^-
Molybdenum	Mo		MoO_4^-
Cobalt	Co	Co^{++}	

*Other anionic forms of phosphorus, sulfur, and boron may be available for plant uptake if soil pH levels are increased. These three elements are also not very susceptible to leaching even though they are in the anionic state.

ABSORPTION OF ELEMENTS BY PLANTS

Generally, there are two mechanisms for the absorption of nutrients by plants.

1. Plant nutrients may be taken directly into the plant as it takes up water from the soil solution. Because of this, it is essential that we regulate the soil solution so that the proper nutrient balance is present.

2. A second and more complicated process is known as the *exchange process*. This process comes about by the plant roots forming carbon dioxide (CO_2) during respiration (an energy generating process). This CO_2 combines with water to form carbonic acid (H_2CO_3). The H_2CO_3 breaks down (ionizes) to form hydrogen ions (H^+) and bicarbonate ions (HCO_3^-). When this happens,

the H^+ in or at the root surface exchanges with nutrients adsorbed on the soil colloid. Schematically, this exchange is depicted in Figure 10-1.

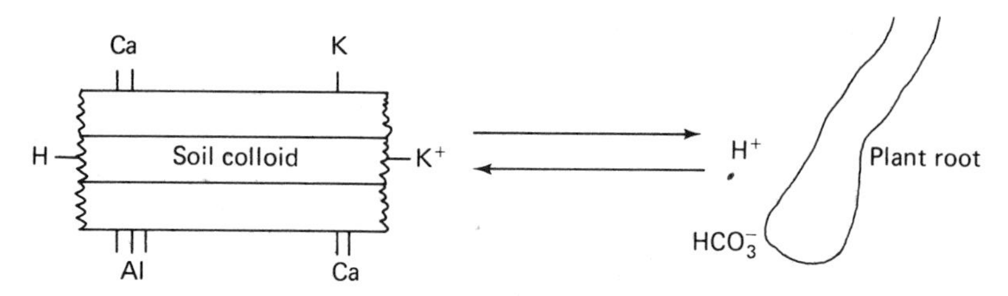

Figure 10-1. Nutrient uptake by cation exchange. The K^+ ion is exchanged with the H^+ ion on the root to achieve potassium uptake.

SOURCES OF ELEMENTS IN THE SOIL

In order to take full advantage of all available nutrients in the soil, a farmer, greenhouse operator, or homeowner must know where all soil nutrients occur. There are actually three major nutrient storage areas in the soil. In these areas can be found all naturally occurring nutrients, as well as many of the applied nutrients. For simplicity, these storage areas are listed below:

ORGANIC MATTER: Most soil nutrients are contained in organic matter. Some persist or remain longer than others. To make many of these nutrients available, the organic matter must be decomposed. The processes affecting this decomposition were outlined in Chapter 9.

SOIL MINERALS (including clay minerals): In Chapter 3 the nutrients contained in various minerals were outlined. These nutrients will become available through weathering. The wise farmer, greenhouse operator, or homeowner will be aware of how fast this process is taking place and how many additional nutrients must be added.

ADSORBED NUTRIENTS: These are the nutrients which are held on the surface or between layers of the soil colloids. This is the major source of nutrients for plant growth and the source most readily controlled by man.

There are usually several nutrients in the soil in unavailable forms because of excessive soil acidity or an improper nutrient balance. These nutrients can often be made more available to plants by adjusting the soil pH.

EXPRESSION OF NUTRIENT CONTENTS OF SOILS, PLANTS, AND FERTILIZERS

Not all soil scientists, agronomists, fertilizer manufacturers, and personnel of soil testing laboratories "speak the same language" when expressing the nutrient content of soils and plants. Most soil scientists, agronomists, and soil testing personnel prefer to discuss soil and plant nutrients as chemical elements. That is, they refer to nitrogen as N, phosphorus as P, potassium as K, calcium as Ca, magnesium as Mg, etc. Sometimes agronomists and nearly all fertilizer manufacturers still refer to some elements, especially phosphorus (P) and potassium (K), in terms of oxides P_2O_5 and K_2O, respectively.

This reference, on an oxide basis, is a holdover from many years ago when most analytical laboratories determined elements as oxides. Today this dual system seems quite confusing. Considerable educational efforts by the American Society of Agronomy, the Soil Science Society of America, and university personnel are under way to change the system to a purely elemental basis. However, the complete conversion is still sometime in the future.

To the beginning student this confusion need not be necessary as long as he remembers that most fertilizers are expressed in terms of actual nitrogen (N), available phosphate (P_2O_5) and available potash (K_2O). What this really means is that in a bag of 10-10-10 there is actually 10 pounds of nitrogen, enough phosphorus to make 10 pounds of P_2O_5, and enough potassium to make 10 pounds of K_2O. There is really only 4.4 pounds of P and 8.3 pounds of K in the bag. On an elemental basis this bag of fertilizer would have been labeled 10-4.4-8.3. To convert from P_2O_5 to P, always multiply by 0.44 and to convert K_2O to K, multiply by 0.83. Conversely, to convert P to P_2O_5 and K to K_2O, multiply by 2.27 and 1.20, respectively.

SUMMARY

This chapter has been presented as an overview of plant growth and nutrition. It is in no way complete. Several good references for additional material follow the Review Questions. It is hoped that the material presented will create an appreciation of the essential nature of plant nutrients, their forms in the soil, and how they are taken up. Discussion of the individual nutrients has been left for later chapters.

REVIEW QUESTIONS

1. What do plants do during the growth processes?
2. What elements are essential for plant growth?
3. What is the difference between macro- and micronutrients?
4. Why must fertilizer be used to produce high yields of crops?
5. In what forms are nutrients taken up by plants?
6. How are nutrients absorbed by plants?
7. Where are nutrients stored in the soil?
8. What is the difference in expressing elements on an oxide versus an elemental basis?

REFERENCES

Brady, Nyle C. *The Nature and Properties of Soils.* 8th ed. New York: Macmillan Co., 1974, pp. 19-38.

Donahue, Roy L., Shickluna, John C., and Robertson, Lynn S. *Soils: an Introduction to Soils and Plant Growth.* 3rd ed. Englewood Cliffs, N.J.: Prentice-Hall, 1971, pp. 222-40.

Foth, H. D., and Turk, L. M. *Fundamentals of Soil Science.* 5th ed. New York: John Wiley, 1972, pp. 13-22, 273-94.

McVickar, Malcolm H. *Using Commercial Fertilizers: Commercial Fertilizers and Crop Production.* 3rd ed. Danville, Ill.: Interstate Printers and Publishers, 1970, pp. 17-35.

Northern, Henry T. *Introductory Plant Science.* 2nd ed. New York: Ronald Press Co., 1958, pp. 74-78, 235-52.

Tisdale, S. L., and Nelson, W. L. *Soil Fertility and Fertilizers.* 3rd ed. New York: Macmillan Co., 1975, pp. 21-65, 66-104.

White, W. C., and Collins, D. C. *The Fertilizer Handbook.* Washington: Fertilizer Institute, 1972, pp. 20-21.

11

Nitrogen, Phosphorus, and Potassium Use by Plants

"Why are the elements nitrogen, phosphorus, and potassium so important?"

"Because they're used in large amounts."

"Because they're controlled by the crop manager."

Because nitrogen, phosphorus, and potassium are used in rather large quantities by plants, it is important that their function in plants and their manner of uptake be given attention. A brief discussion of these nutrients follows.

NITROGEN

This nutrient element is probably the most important of any we will study. Nitrogen has sometimes been called the *growth element* primarily because it is a vital part of plant protoplasm. Protoplasm is the seat of cell division and, therefore, plant growth.

Nitrogen, unlike most other fertilizer elements, is not found to any great extent in soil-forming rocks and minerals. The source of practically all soil nitrogen is the atmosphere. Fortunately, it contains a nearly unlimited supply of nitrogen. Approximately four-fifths (80 percent) of the air we breathe is nitrogen.

Although an abundance of nitrogen (N_2) gas occurs over every acre of the earth's surface (approximately 37,000 tons per acre of surface), it cannot be utilized by growing plants in the elemental (N_2) form. It must be combined with other elements (primarily oxygen or hydrogen) before it can be used by plants.

Nitrogen is supplied to the soil both from the atmosphere and from fertilizer. Sources from fertilizer will be discussed in Chapter 13.

Sources of Soil Nitrogen

ORGANIC MATTER: Organic matter may be considered as the soil's storehouse of nitrogen. As previously indicated, organic matter is the temporary end product in decomposition of plant and animal residues. The nitrogen content of organic matter varies, but averages from 4-5 percent nitrogen. Therefore, a soil with 4 percent organic matter contains about 4,000 pounds of reserve nitrogen per acre per 6 inches in depth. Not all of this nitrogen is available to growing plants in one season.

Humus, the end product of organic matter breakdown, or of plant and animal residue decomposition, is relatively stable in the soil. Before plants can utilize the nitrogen in humus, it must be further decomposed. Decomposition of humus progresses at a relatively slow rate. Consequently, the amount of nitrogen available in any one growing season through humus decomposition is a small percent of the total amount in the soil. Man may speed up the release of nitrogen from humus in some soils by liming to a proper pH, draining of flooded or excessively wet soils, or in some cases even aerating by excessive tillage.

On the mineral soils of the southeastern United States, not more than 15 to 25 pounds yearly of organic nitrogen are available for plant growth. Much higher amounts are available from the prairie soils of the Midwest. The exact amounts will depend on the organic matter contents and the yearly organic matter additions. The soils of the northeastern United States may contribute slightly more than the Southeastern soils particularly if they have recently grown legumes.

SYMBIOTIC SOIL ORGANISMS: The major portion of elemental nitrogen that finds its way into the soil for plant use comes from soil atmospheric nitrogen fixed by special purpose bacteria attached to the roots of legume plants such as alfalfa, clover, soybeans, peanuts, etc. Amounts ranging from 40 to 120 pounds of nitrogen per acre per year may be fixed by this mechanism.

NONSYMBIOTIC SOIL ORGANISMS: Certain types of nonsymbiotic bacteria (not in association with legume roots) have the ability to fix atmospheric nitrogen. Since these organisms require an energy source high in carbon, the amount of nitrogen fixed depends on the amount of organic matter the soil contains. Also soil pH determines the amount fixed. Azotobacter, one of the major strains of nonsymbiotic nitrogen-fixing bacteria, functions most effectively at a pH of 6.0-7.5. Usually 10 to 25 pounds of nitrogen per acre per year may be fixed by nonsymbiotic organisms.

RAINS AND SNOW: Electrical discharges from lightning generates heat that causes atmospheric oxygen to react with nitrogen to form several kinds of nitrogen oxides. Precipitation brings these oxides to the soil surface. These chemically active oxides combine with other elements to form salts, such as sodium nitrate ($NaNO_3$) and/or potassium nitrate (KNO_3). Plants absorb the nitrate nitrogen from these salts. Approximately five pounds of nitrogen per acre per year is added to the soil by rainfall and snow.

Forms of Nitrogen Used by Plants

Plants take up both ammonium (NH_4^+) and nitrate (NO_3^-). After taking up these two forms of nitrogen, the plants convert them to amino acids and then to the more complex protein molecule. If the plant takes up NH_4^+ during cool weather when growth is slow, it may have to use its own stored carbohydrates for this conversion. If it takes up NO_3 during this period, it can store it until a more favorable time for conversion. Because of these conditions, it is wise to apply nitrate nitrogen to crops growing in cool seasons.

The source of nitrogen isn't as critical when applied to crops growing during warm seasons of the year because the nitrifying bacteria will convert ammonium to nitrate (nitrification). Some crops such as potatoes, tomatoes, and tobacco don't respond as well to ammonium nitrogen as nitrate nitrogen.

Functions of Nitrogen in Plants

Regardless of the form absorbed by plants, it is converted within the plant to the reduced or NH_2 form. This reduced nitrogen is combined with organic acids to form amino acids. Amino acids are considered the building blocks of proteins, which in turn are used in forming cell protoplasm, the seat of many plant processes. Ultimately, nitrogen controls plant growth and reproduction. A nitrogen deficiency slows down protein formation and, ultimately, growth.

One of the most noticeable effects of nitrogen, especially if abundantly supplied to plants, is the dark green color that develops in the leaves. Nitrogen is essential in chlorophyll formation (the green coloring material in leaves). By way of contrast, plant nitrogen deficiency results in a yellowing of the leaves. Nitrogen and magnesium are the only two soil supplied elements that are found in the chlorophyll molecule. The presence of chlorophyll is necessary for carbon, hydrogen, and oxygen to be converted into simple sugars. These sugars and their conversion products are essential to plant growth and development.

Nitrogen Balance in the Soil and Atmosphere

It is common to hear soil scientists refer to *nitrogen balance* in the soil. This term really means the balance between nitrogen in the atmosphere, that removed by crops, the losses due to volatilization and drainage, and the nitrogen contributed by organic materials, fertilizers, and nitrogen-fixing bacteria. A somewhat simplified nitrogen cycle is shown in Figure 11-1.

The nitrogen cycle tells us what can happen to various forms

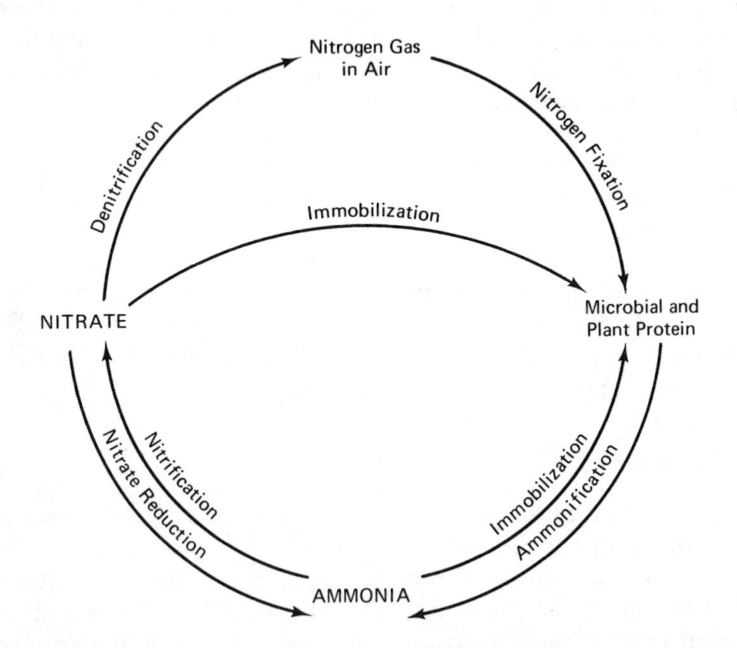

Figure 11-1. The nitrogen cycle. (Adapted from F.E. Broadbent, "Organic Matter," in *Soil, 1957 Yearbook of Agriculture.* Washington: USDA, United States Government Printing Office, 1957, p. 154.)

of nitrogen and the transformations they undergo as they cycle from the atmosphere to the soil and back.

PHOSPHORUS

The element phosphorus was discovered in the 1600s, but it was nearly two hundred years later that it was found to be essential for plant growth. Many soils of eastern United States are deficient in available phosphorus and plants therefore respond to its application. On the average the content of native phosphorus in surface soils is only about one-half that of nitrogen and one-twentieth that of potassium.

In most soils native soil phosphorus is bound or fixed in insoluble forms such that only a very small part of the total supply becomes available in any one cropping season. For this reason, even soils with a high total phosphorus content often fail to provide sufficient available phosphorus for maximum crop production.

Soil phosphorus comes from the decomposition of rocks containing the mineral apatite ($Ca_{10}(PO_4)_6(F, Cl, OH)_2$). (See Chapter 3.) This mineral is also the major source of all commercial phosphate fertilizers today. Organic matter is a source of soil phosphorus. However, it generally accounts for a small percentage of most available soil phosphorus.

Importance of Phosphorus in Plant Growth

Except for nitrogen, unsatisfactory plant growth more often is due to a shortage of this element than of any other. Phosphorus is intimately associated with all life processes and is a vital constituent of every living cell. This element tends to be concentrated in the seed. Phosphorus has been assigned the following functions in growth and reproduction of plants.

1. Stimulates early root formation and growth.
2. Gives a rapid and vigorous start to plants.
3. Hastens maturity.
4. Stimulates blooming and aids in seed formation.
5. Gives winter hardiness to fall-seeded grains and hay crops.
6. Is extremely important to germinating seedlings.

The phosphorus uptake of plants follows the growth of plants quite closely. On a weight basis, young plants have a higher percentage of phosphorus than older plants. By the time young plants have reached about 25 percent of their total dry weight, they will have accumulated as much as 75 percent of their total phosphorus. The phosphorus content of plants is considerably smaller than nitrogen or potassium, e.g., a cotton plant may have only one-tenth as much phosphorus in its tissue as it has nitrogen content, both when an early developing seedling, as well as when it approaches maturity.

Forms of Phosphorus in the Soil

On the generally acid soils of the eastern and southeastern United States, phosphorus usually occurs and is taken up by the plants as primary orthophosphate ($H_2PO_2^-$). On more alkaline soils (pH more nearly 7 or even above) it may occur and be taken up as secondary orthophosphate ($HPO_4^=$). However, the primary orthophosphate ($H_2PO_4^=$) is the major form. Because the orthophosphate is an anion, it can enter into chemical reactions to form compounds with calcium (Ca^{++}). iron (Fe^{+++}), and aluminum (Al^{+++}). When this happens, the phosphate is no longer available to plants. This reaction is called *phosphorus fixation.* The type of reaction (chemical combination) will depend on soil pH. If the soil is acid ($pH<6.0$), iron and aluminum compounds will occur as:

$$Al \left\langle \begin{array}{l} OH \\ OH \\ H_2PO_4 \end{array} \right. \qquad \text{or} \qquad Fe \left\langle \begin{array}{l} OH \\ OH \\ H_2PO_4 \end{array} \right.$$

If the soil is alkaline, the phosphorus will combine with calcium to form apatite ($Ca_5(PO_4)_3F$). The only way to slow this process of fixation is by holding the soil pH at about 5.8-6.5.

Phosphorus is also tied up by clay and organic matter. These fixation mechanisms are quite detailed and will not be discussed in this text. If phosphorus is fixed by clay, it may be possible to satisfy the clay's fixing power by adding large applications of phosphorus to the soil.

Phosphorus may be tied up in the metabolism and growth of the soil organisms, thus temporarily immobilizing it. The phosphorus in these forms is eventually returned to the soil, and becomes available to growing plants on the death and decay of the soil organisms. Generally this process accounts for a small percent of the phosphorus fixed in soils.

Management Considerations to Improve Phosphorus Utilization

Because phosphorus is so readily fixed in many soils, consideration should be given to decreasing fixation and increasing utilization. The following lists suggest certain practices which may do this.

1. Adjust the soil to a desirable pH through proper liming programs. This will reduce iron and aluminum fixation in acid soils.
2. Strive to maintain or increase organic matter levels to keep up the release of organic phosphorus.
3. Use band applications at planting time. This will delay the mixing of phosphorus with the soil.
4. Broadcast a large amount of phosphorus to satisfy to some extent the fixing power of the soil.

Soils throughout the eastern United States vary in their phosphorus-fixing capacities. The exact fixing capacity will depend on the soil pH and the aluminum, iron, and calcium contents of the soils. Generally, the red and yellow clay soils of the Southeast have high to very high fixing capacities, while the sandy coastal plain soils only fix low to medium quantities. In the northern and midwestern United States, the loess and glaciated soils, with pHs between 6 and 7, have low fixing capacities. Calcareous (high calcium carbonate) glacial till and loess soils have relatively high fixing capacities due to the calcium present. The acid organic soils of the Southeast have very high fixing capacities due to high aluminum contents. Generally the alluvial soils along the Mississippi River are low phosphate fixers.

POTASSIUM

Potassium occurs as a simple cation throughout the soil-plant-system. It is rather mobile, and therefore, subject to considerable leaching in sandy soils, especially if the pH is relatively low. Potassium occurs on the exchange complex of all soils but to a much greater extent on clayey soils.

Of the three macronutrients used in greatest amount in plant growth, potassium is the most abundant in the earth's crust. However, soils vary greatly in their total potassium content. Soil potassium content is related to the material from which the soil was formed and the degree of weathering. Highly weathered soils derived from rocks low in potassium generally contain the lowest amount of the element.

Importance of Potassium in Plant Growth

Although not completely understood, there are several plant functions ascribed to this element. Unlike nitrogen and phosphorus, potassium apparently does not occur as part of organic compounds formed in the plant. It is extremely mobile in the plant, and is readily translocated from one part of the plant to another.

The main function of potassium appears to be associated with plant metabolism. It is required for photosynthesis, and when deficient, a reduction in this process occurs. A deficiency of potassium also seems to produce an increased respiration rate. Potassium seems to be essential in protein synthesis because potassium deficient plants seem to be unable to synthesize protein.

An adequacy of potassium seems to reduce lodging of tall growing crops such as corn, small grain, and sorghums. These crops and others when receiving high rates of nitrogen, coupled with inadequate potassium, are more prone to lodging. This nutrient is also important in water economy of plants. Adequate amounts promote turgidity (rigidity due to adequate water in leaf cells), and thereby maintain the internal pressure of plant tissue. Finally, fruit and vegetable quality are frequently associated with plant potassium. Fruits and vegetables from deficient plants are usually small, misshapen, poor in color, and subject to rot quickly in storage.

Forms of Potassium in the Soil

There are three types of potassium occurring in the soil: *unavailable*, *slowly available*, and *readily available*. Unavailable potassium is the reserve potassium that is contained in the structure of micas and feldspars. This form is of minimum importance in the nutrition of plants.

Slowly available is that potassium which is temporarily attached along the edges of the interlayer spaces of illite, vermiculite, and montmorillonite. This type of attraction or retention is confined mostly to 2:1 type clays, much more common in the northern and midwestern sections of the United States. These soils have clay colloids which expand when wet and contract when dry. It has been suggested the soil potassium can be trapped and held between the silicon and aluminum layers of these soils during the expansion and contraction process. If held in this manner it is not easily released and, consequently, it is very slowly available to growing plants. Fixation of soil potassium to slowly available forms is of major concern in midwestern areas of the United States, where montmorillonitic-types of soil clays predominate. It is of less concern on the more highly weathered kaolinitic soils of the southern and southeastern areas of the United States.

In the South and portions of eastern United States, the most prevalent form is the readily available. Only some very young soils of the mountainous regions contain the slowly and unavailable forms of potassium. The readily available form of potassium is contained in the soil solution and/or adsorbed by soil colloids. Since potassium is a positively charged ion (K^+), it is readily held. In this form it is considered readily available to plant roots.

The readily available forms of soil potassium are the major sources absorbed by plants and measured in soil analyses for determining if there is a need for supplementing the soil with fertilizer potassium. This kind of potassium may easily be lost from the soil through leaching. Consequently, the supply of available potassium can be increased by potassium fertilization. The more clay or organic matter a soil contains, the greater its ability to store this element when it is applied as fertilizers.

Losses of Potassium from the Soil

Because this element is so mobile and seldom becomes associated in complex molecules of starch, proteins, or fatty substances, it is always subject to loss from the soil-plant system. For simplicity, these methods of loss are listed as follows:

CROP REMOVAL: Most crops remove more potassium from the soil than any other plant nutrient element, except nitrogen. Under conditions of heavy nitrogen fertilization and removal of all the forage, potassium removal may exceed nitrogen removal. Some kinds of plants (forage grasses and legumes, especially) will continue to absorb this potassium above their normal needs. This is called *luxury consumption*, and is frequent under heavy potassium fertilization. When this happens, there has been uneconomical utilization of fertilizer potassium. The likelihood of this happening can be avoided by applying only recommended amounts, as determined by soil analyses, and in some instances, split applications of potassium fertilizers during the growing season. High nitrogen rates and inadequate potassium fertilization can result in rapid potassium depletion on some soils, particularly where deep, sandy surfaces occur. This is particularly true for forage crops where all or most of the forage is removed.

LEACHING OF POTASSIUM: On deep, sandy soils considerable fertilizer potassium may leach below the root zone of growing plants. This leaching may occur to a greater extent if the soil pH is below optimum. On clayey soils sufficient potassium may occur, or if added in the fertilizer, will persist throughout that season or even carry over for the second crop.

The leaching of potassium can be reduced by proper liming. It was shown in Chapter 6 that the surface of clay particles and organic colloids are covered with many negative (–) charges. These negatively charged particles attract positively charged ions including potassium. Due to ion size and charge characteristics, elements differ in their ability to attach to clay and organic particles in the following decreasing order: aluminum (Al^{+++}), hydrogen (H^+), calcium (Ca^{++}), magnesium (Mg^{++}), potassium (K^+), ammonium (NH_4^+), and sodium (Na^+). Because K^+ is held less tightly than some of the above elements, it gets pushed out into the downward moving soil water and into drainageways. It is easier for potash to stay on the clay particle when it is competing with calcium (Ca^{++}) and magnesium (Mg^{++}), on a well-limed soil, than when it is competing with a lot of aluminum (Al^{+++}) and hydrogen (H^+) on acid soils. Therefore, decreased leaching of potash is an important reason for adequate liming of soils.

EROSION: Any time that soil is carried away from the surface of fields by water and wind erosion, there will be a loss of absorbed potassium or minerals that contain this necessary nutrient. Although much of the potassium that is lost is not readily available, it is still a potential and valuable nutrient loss.

SUMMARY

This section was presented to give the student a brief background of the functions of nitrogen, phosphorus, and potassium in plants. Information about these nutrients was presented because these elements are of prime importance in management efforts to achieve a high level of soil fertility and to enhance crop yields.

REVIEW QUESTIONS

1. What forms of nitrogen, phosphorus, and potassium are taken up by plants?

2. What are some of the most important and well-known functions of these elements in plant growth?

3. What are the important concepts stressed in the nitrogen cycle?

4. What is phosphorus fixation?

5. How can phosphorus utilization be improved?

6. Define the terms unavailable, slowly available, and readily available potassium.

7. How is potassium lost from soil?

8. Why does proper liming reduce the amount of potassium leached from soils?

9. What is luxury consumption of potassium?

REFERENCES

Brady, Nyle C. *The Nature and Properties of Soils*. 8th ed. New York: Macmillan Co., 1974, pp. 422-44, 456-83.

Donahue, Roy L., Shickluna, John C., and Robertson, Lynn S. *Soils: an Introduction to Soils and Plant Growth*. 3rd ed. Englewood Cliffs, N.J.: Prentice-Hall, 1971, pp. 227-30.

Foth, H. D., and Turk, L. M. *Fundamentals of Soil Science*. 5th ed. New York: John Wiley, 1972, pp. 307-13.

Kilmer, V. J., Younts, S. E., and Brady, Nyle C. *The Role of Potassium in Agriculture*. Madison: American Society of Agronomy, 1968, pp. 23-33, 79-94.

McVickar, Malcolm H., *Using Commercial Fertilizers: Commercial Fertilizers and Crop Production*. 3rd ed. Danville, Ill.: Interstate Printers and Publishers, 1970, pp. 17-35.

Olson, R. A., et al., *Fertilizer Technology and Use*. 2nd ed. Madison: Soil Science Society of America, 1971, pp. 217-51, 285-96.

Tisdale, S. L., and Nelson, W. L. *Soil Fertility and Fertilizers*. 3rd ed. New York: Macmillan Co., 1975, pp. 127-88, 208-42, and 243-61.

White, W. C., and Collins, D. C. *The Fertilizer Handbook*. Washington: Fertilizer Institute, 1972, pp. 16-17, 30-39, and 56-63.

12

Sulfur, Magnesium, Calcium, and Trace Elements

"What are secondary nutrients and trace elements?"

"Secondary nutrients are the macronutrients —sulfur, magnesium, and calcium."

"Trace elements are micronutrients, or nutrients used in small quantities."

"Remember! They're all *very important."*

As indicated in Chapter 10, sulfur, magnesium, and calcium are macronutrients that on occasion are referred to as *secondary* elements. This designation is a holdover from a time when agronomists were not as concerned about supplying these nutrients for crop production as they were about supplying nitrogen, phosphorus, and potassium. At that time it was thought that most soils contained sufficient natural supplies; or commercial fertilizers supplied considerable quantities of these elements as impurities and carriers of nitrogen, phosphorus, and potash.

Today's fertilizers that supply nitrogen, phosphorus, and potassium are refined so that the secondary elements are reduced or removed. Also, intensive cropping and greater emphasis on higher yields per acre have increased the need for all essential elements.

SULFUR

Function in Plants

Although crops vary in their sulfur requirements, most plants require about the same amount of elemental sulfur as they do of elemental phosphorus. Sulfur is important in the development of amino acids, the building blocks of plant protein. It is also necessary for increased root growth, maintenance of dark green color, promotion of nodule formation on legumes, stimulation of seed production, and the encouragement of more vigorous plant growth. The maintenance of dark green color is related to chlorophyll development. Although sulfur is not a constituent of chlorophyll, plants deficient in this element exhibit a pale green or yellow color which generally appears first on the upper younger leaves.

Forms of Sulfur

Sulfur is taken up by plants mostly as the sulfate ion ($SO_4^{=}$). Since it is an anion, it is not readily adsorbed by soil colloids. Thus, it can leach through the soil profile similar to the nitrate (NO_3^-) ion. Like the phosphate ion ($H_2PO_4^-$), it is held to some degree by the clays and hydrous oxides of soils.

Sources of Sulfur

Sulfur is supplied both by the organic fractions of the soil and the atmosphere, as well as from commercial fertilizers.

a. Soil and atmosphere—considerable quantities of sulfur are tied up (or released upon further decay) in soil organic matter. In most soils, organic matter is the major source of soil sulfur. Organic matter must first be decomposed before the sulfur is released for use by plants. Soil sulfur is replenished by the gas sulfur dioxide (SO_2) which is contained in the atmosphere. It is brought down to the soil by rainwater and snow. Sulfur dioxide is also furnished by sulfur containing fertilizers and by insecticides.

The amount of sulfur added to soils from sulfur dioxide in the atmosphere varies considerably from one area to another. Areas near industrial plants and urban centers have the greatest atmospheric concentration of sulfur dioxide. Rural areas, characterized by farms and small communities, have the smallest atmospheric contribution. For example, in North Carolina, measurements in the early 1950s showed concentrations of elemental sulfur (S) brought down annually in precipitation to range from 5 to 16 pounds per acre. In some highly industrialized areas of northern Indiana, parts of Ohio, Pennsylvania, and New York, as much as 40 pounds of sulfur (S) are washed out of the atmosphere and enter the soil. Where natural gas and other petroleum products have replaced coal as a source of fuel, and more stringent pollution systems on industrial plants have been implemented, less sulfur is being returned to the atmosphere in the form of sulfur dioxide. This fact, along with increased use of high analysis fertilizers containing little or no sulfur, may lead to more sulfur deficiencies in the future unless supplementary sources of sulfur are used by farmers.

b. Fertilizer—there are numerous products available for use by the fertilizer industry, farmers, and horticulturists for supplying sulfur to soils and crops. Many fertilizer materials commonly used in producing mixed fertilizer, or used as separate materials, contain considerable quantities of sulfur. Even elemental sulfur, if it is ground sufficiently fine, is a good source. Table 12-1 lists some common sources of sulfur.

Table 12-1
Fertilizer Sources of Sulfur

Source	*Percent S*
Ammonium sulfate $(NH_4)_2SO_4$	24
Copper sulfate ("bluestone") $CuSO_4$	13
Epsom Salt $MgSO_4 \cdot 7H_2O$	14
Ferrous sulfate (copperas) $FeSO_4$	12
Gypsum $CaSO_4 \cdot 2H_2O$	17
Manganese sulfate $MnSO_4$	14.5
Normal superphosphate $Ca(H_2PO_4)_2 + CaSO_4 \cdot 2H_2O$	12
Potassium-magnesium sulfate $K_2SO_4 \cdot 2MgSO_4$	22
Elemental sulfur S	30-100
Triple superphosphate $Ca(H_2PO_4)_2$	2
Zinc sulfate $ZnSO_4 \cdot H_2O$	18

Reprinted in part by permission of *The Fertilizer Handbook*, 1972, a publication of the Fertilizer Institute, Washington, D.C., p. 71.

MAGNESIUM

Function in Plants

Magnesium is an essential element which is normally absorbed as the magnesium ion (Mg^{++}). Chlorophyll, which is necessary for photosynthesis, contains 2.7 percent magnesium. In other words, magnesium helps give the green color to leaves. Magnesium is also necessary in regulating uptake of other plant foods: it acts as a carrier of phosphorus in the plant, it promotes the formation of oils and fats, and seems to play a part in the translocation of starch.

Reaction with Soils

Soil supplied magnesium comes largely from the decomposition of rocks containing such minerals as biotite, hornblende, dolomite, chlorite, serpentine, olivine, etc. (Chapter 3). Since magnesium is a positively charged cation, it may be adsorbed by the soil colloidal fraction or remain present in the soil solution. In this form it is readily available to growing crops.

On soils with deep, coarse-textured surfaces and limited nutrient retention capacity, magnesium fertilization management must be considered. This situation would be generally true of much of the Atlantic Coastal Plain and other areas in the southeastern United States where low soil pH is more often the rule than the exception. These soils are inherently low in magnesium. Consequently, magnesium deficiency in crop production is more likely to occur. The least expensive and most practical method of correcting a possible deficiency is to use dolomitic limestone (a mixture of calcium and magnesium carbonate) as a liming material. On high pH soils where a quick response to magnesium is desired, a more soluble source of magnesium should be used.

Source of Magnesium

Soil magnesium sources were mentioned above as coming from weathering of rocks and minerals. Sources of materials used in fertilizer manufacture or available to producers for direct application are listed in Table 12-2.

Table 12-2
Fertilizer Sources of Magnesium

Source	*Percent Mg*
Dolomitic limestone $MgCO_3 + CaCO_3$	Variable
Epsom salt $MgSO_4.7H_2O$	9.6
Kieserite, calcined $MgSO_4.H_2O$	18.3
Magnesia MgO	55
Potassium-magnesium sulfate $K_2SO_4.2MgSO_4$	11.2

Reprinted in part by permission from *The Fertilizer Handbook,* 1972 a publication of the Fertilizer Institute, Washington, D.C., p.69.

CALCIUM

Function in Plants

Plants absorb calcium as the ion (Ca^{++}). Calcium ions can come from the soil solution or by actual contact of the root with the clay particle holding the calcium. This nutrient ion promotes early root formation and growth, improves general plant vigor and stiffness of straw, helps neutralize undesirable organic acids within the plant, encourages grain and seed production, and increases the calcium content of food and feed crops. Calcium deficiencies, as such, are actually a relatively rare occurrence in plants. Some examples of deficiency symptoms are "pops" in peanuts (empty peanut shells because the fruit fails to develop), occasionally the failure of new emerging corn leaves to unfold as if stuck together, blossom-end rot in tomatoes, and abnormal tobacco leaf development (severely twisted or crinkled). All of these problems can be corrected by the proper calcium materials.

Reactions Within the Soil

Calcium ions may be lost in drainage water, absorbed by soil organisms and temporarily immobilized, adsorbed and held by soil colloids, or reprecipitated as secondary calcium compounds in low rainfall climates. Before being adsorbed, calcium ions must first replace other adsorbed cations on the exchange sites of clay and organic matter. Since most soils of eastern and southeastern United States are acid, hydrogen or aluminum is generally the adsorbed cation replaced.

Amounts of Soil Calcium Needed for Plant Growth

The amount of calcium needed in the soil for optimum plant growth is related to the ability of the clay and organic matter to hold cations (cation exchange capacity). This fact is illustrated in Table 12-3.

The significant point from Table 12-3 is that for optimum

Table 12-3
Calcium Requirements for Optimum Plant Growth on Different Soil Materials

Type of Soil	*Cation Exchange Capacity meq/100 gr*	*Percent Calcium Saturation Needed to be Adequate**	*Pounds Calcium Needed per Acre*
Organic	100	25-40	2,500-4,000
Kaolinitic	8	40-60	300-400
Montmorillonitic	10	60-80	600-800
Sandy	2	50	400

*Percentage of the cation exchange capacity that needs to be occupied by calcium to insure good plant growth.

plant growth it takes 2,500 to 4,000 of calcium per acre on high organic soils, and only 400 pounds per acre on low organic soils with deep, sandy surfaces. Soil tests must be calibrated (or adapted) against actual field conditions based on field research to determine this information. These circumstances emphasize the need for having soils tested by a laboratory using locally adapted tests for locally grown crops. It also emphasizes that pH is only one factor in determining an adequate liming program.

Sources of Calcium

The major sources of calcium are given in Table 12-4. The source chosen will depend on several factors including price, need for other nutrients, and the effect the material will have on soil pH.

Table 12-4
Sources of Calcium

Source		*Percent Ca*
Blast furnace slag	$CaSiO_3$	29
Burned lime	CaO	60
Calcitic limestone	$CaCO_3$	32
Dolomitic limestone	$CaCO_3$ $MgCO_3$	Variable
Hydrated lime	$Ca(OH)_2$	46
Marl	$CaCO_3$	20-75
Calcium cyanamide	$CaCN_2$	38
Calcium nitrate	$Ca(NO_3)_2$	19
Phosphate rock	$3Ca_3(PO_4)_2.CaF_2$	32-34
Superphosphate, normal	$Ca(H_2PO_4)_2$	20
Superphosphate, triple	$Ca(H_2PO_4)_2$	14
Gypsum	$CaSO_4$	29

Reprinted in part by permission from *The Fertilizer Handbook,* 1972, a publication of the Fertilizer Institute, Washington, D.C., p. 68.

RELATIONSHIPS AMONG MAGNESIUM, CALCIUM, AND POTASSIUM

A significant relationship exists between magnesium, calcium, and potassium in the soil; this condition is demonstrated in the availability of these nutrients to plants. These three nutrient ions, and to some extent even sodium (although not an essential element), occupy exchange positions or sites on soil colloids. An overabundance of any one or two of these ions may cause them to exchange with the others and accelerate leaching of the exchanged ions to below the plants' effective root zone.

For example, a magnesium deficiency may be accentuated by an imbalance between calcium and magnesium in the soil. If the ratio of calcium to magnesium in the soil is too high, magnesium absorption by plants may be depressed. This situation may occur when calcitic lime has been the only liming material used for a number of years on soils relatively low in magnesium. This condition may also occur in a crop following peanuts where excessively high rates of gypsum have been applied. Also, much residual calcium (unused by peanuts) may interfere with adequate uptake of magnesium by the following crop. Sometimes excessive fertilizer potassium applications may interfere with sufficient plant uptake of magnesium. It has even been shown that very high soil magnesium levels from sustained use of dolomitic lime or excessive fertilizer magnesium applications may interfere with potassium or calcium uptake.

Soil testing laboratories generally recognize soils that have serious imbalances of these nutrient ions and offer suggestions to alleviate this problem. A general rule to remember is that for each available magnesium ion there should be about six calcium ions for satisfactory crop production.

MICRONUTRIENTS (MINOR OR TRACE ELEMENTS)

Micronutrients, *minor elements*, or *trace elements* are the names given to a group of seven elements that are needed in very small amounts. These trace elements include boron (B), manganese (Mn), zinc (Zn), iron (Fe), copper (Cu), molybdenum (Mo), and chlorine (Cl). All except boron and chlorine are absorbed by plants as positively charged ions. Boron is absorbed as the anion ($B_4O_7^{--}$) and chlorine as (Cl^-). It is difficult to make more than just a few general statements about this group of seven elements. They react differently with different soil components, are influenced differently by soil moisture conditions, and are needed in different amounts depending on plant species. The plant availability of these elements is generally reduced by an increase in soil pH. Specifically the availability of boron, copper, iron, manganese, and zinc decreases as pH increases from 5 to 7. The reverse is true for molybdenum: its availability increases as pH increases. These relationships are shown in Figure 12-1. It can be readily noted from Figure 12-1 that micronutrient availability is regulated by soil acidity. Therefore man, through management of his liming program can strongly influence the supply of plant-available micronutrients. An excess of a micronutrient may occur on acid soils (iron) or a deficiency occur on a well-limed soil (manganese, zinc, etc.). On many soils with a pH falling below 5.5 there may be an excessive accumulation of iron, which may accumulate in the nodes and block the translocation tissues of the plant. Molybdenum is

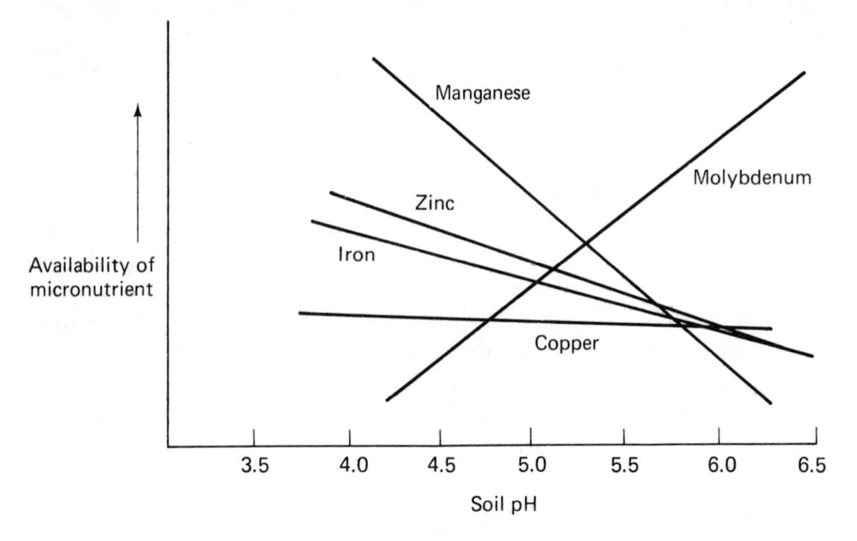

Figure 12-1. Effect of soil pH on micronutrient availability. (Developed from the authors' personal communications with Dr. F.R. Cox, Professor of Soil Science, North Carolina State University, Raleigh, North Carolina.)

likely to be deficient and is frequently supplied as a seed treatment on peanuts, soybeans, and forage legumes as an insurance factor. In addition, aluminum may be in excess. It can form insoluble phosphates (from phosphate fertilizer added) that are unavailable to plants.

In the pH range of approximately 5.5 to 6.2, deficienceis of micronutrients are least likely to occur. Likewise, this is the most optimum pH range for many important crops. Alfalfa will need a higher pH for optimum growth. When the soil pH begins to occur much above 6.2, manganese, zinc, copper, and boron (iron for some ornamentals) are likely to be needed on some crops, especially when the pH reaches 6.5 to 7.0. At these pH levels, manganese will likely be needed by soybeans, peanuts, oats, and cotton.

Specific Micronutrients

BORON: Deficiencies have been reported in over 40 states, with many different crops affected. Generally boron enhances or increases cell division, viability of pollen grains, fruit, seed, or fiber formation, carbohydrate and water metabolism, and protein synthesis. Some deficiency symptoms of selected crops are as follows:

Alfalfa–Stunted growth, bright yellow upper leaves, and reduced seed yields.

Apples–External and internal cork (hard, wrinkled tissue).

Peaches–Dieback of terminal twigs, curled leaf edges, dead buds.

Citrus–Heavy fruit shedding, yellowing of leaf veins.

Cotton–Excessive shedding of flower buds and bolls.

Celery–Cracks crosswise of stalks, "Crack stem."

Peanuts–Deformed kernels, dark spots in kernels.

The soluble form of boron in soils, an anion, is easily leached. In humid areas, coarse-textured soils usually contain the smallest available quantities of this element for alfalfa, corn, soybean, peanut, and vegetable production. Low soil moisture levels reduce boron availability. For this reason, boron deficiency, especially in alfalfa, peanuts, and cotton, is most frequent under drought conditions. Table 12-5 lists selected boron sources.

Application rates are usually in the range of 0.2 to 3 pounds of boron per acre, with the lower rates for the most sensitive crops; beans (soybeans, field beans, edible beans), peanuts, corn, and cotton. Deficient alfalfa and sugar beets require higher rates. Generally a soil application is preferred to foliar sprays except for crops which have routine spray programs. Toxicity can be a serious problem when boron is applied excessively. For example, application rates for sugar beets can cause corn and bean crop failures. Analyses of soils, plants, and irrigation water are the best means of determining boron levels and needs.

Table 12-5
Sources of Boron

Material		*Percent B*
Boron fruits		10-17
Borax	$Na_2B_4O_7.10H_2O$	11
Boric acid	H_3BO_3	17
Sodium tetraborate		
Fertilizer borate-48	$Na_2B_4O_7.5H_2O$	15
Fertilizer borate-68	$Na_2B_4O_7$	21
Solubor	$Na_2B_{10}O_{16}.10H_2O$	20

Reprinted in part by permission from *The Fertilizer Handbook,* 1972, a publication of the Fertilizer Institute, Washington, D.C., p. 73.

COPPER: Deficiencies of copper have been reported in approximately fifteen states. Most often cases of copper deficiency occur in peat and muck soils. Although required in small quantities it is important in chlorophyll formation and in several plant enzyme systems. The reactions of copper in soil are very complex. Deficiencies may occur in acid sandy soils where copper has been leached. Deficiency symptoms for this element are not very specific or unique. Some of the better established ones are as follows:

Citrus—Dieback of twigs, yellow leaves.

Small grains—Twisted leaves, yellowing along leaf margins with "tip burn."

Corn—Yellowing between leaf veins.

Vegetables—Dieback of leaves.

Soil pH influences copper availability to plants, although not as strongly as it does for manganese or to some extent for zinc. Generally, less copper is available in a soil as its pH is raised. This relationship is probably more profound in light, sandy soils low in organic matter.

If too much copper, either in applied fertilizers or in fungicides, has been applied to soils, especially if they are sandy, a toxicity problem may develop. This condition apparently has developed in some Florida citrus groves where as much as 200 pounds of this element has been observed in the soil.

In the high organic regions of the South Atlantic Coastal Plain response to copper has been obtained on corn, soybeans, and small grains. These responses have been frequent enough so that widespread use of supplemental copper has been suggested for all newly cleared land that is brought into crop production. In addition, responses to copper by wheat and occasionally soybeans has been noted on some of the sandy middle Coastal Plains soils along the south Atlantic coast.

Selected copper sources are given in Table 12-6. There are other copper formulations that are not listed in Table 12-6. Since complications often arise in mixing copper materials with fertilizers, instruction from the manufacturer of special-formulation copper carriers should be carefully followed.

Table 12-6
Sources of Copper

Material		*Percent Cu*
Basic copper sulfate	$CuSO_4.H_2O$	13-53
Copper(ic) ammonium phosphate	$Cu(NH_4)PO_4.H_2O$	32
Copper chelates		9-13
Copper chloride	$CuCl_2$	17
Cupric oxide	CuO	75
Cuprous oxide	Cu_2O	89

Reprinted in part by permission from *The Fertilizer Handbook,* 1972, a publication of the Fertilizer Institute, Washington, D.C., p. 75.

IRON: Elemental iron is not often needed in the eastern United States. It is more often needed in the arid regions of the West. Fruit trees and some kinds of ornamentals are the most affected crops. Low levels of available iron contribute to decreased plant chlorophyll formation and impaired respiration processes. Yellowing of tissue between the veins of the younger leaves is a typical symptom of iron deficiencies.

In the more humid areas of eastern United States where soils may be quite acid, iron deficiencies are rare. Sometimes ornamentals, such as azaleas and camelias, may need supplemental iron, especially if they are growing on soils which have had excessive lime applied. In other rather specialized situations, such as citrus in Florida, iron deficiency may be present but is not a widespread problem. Some selected iron materials are given in Table 12-7.

Table 12-7
Sources of Iron

Material		*Percent Fe*
Ferrous ammonium phosphate	$Fe(NH_4)PO_4 . H_2O$	29
Iron ammonium polyphosphate	$Fe(NH_4)HP_2O_7$	22
Ferric sulfate	$Fe_2(SO_4)_3 . 4H_2O$	23
Ferrous sulfate	$FeSO_4 . 7H_2O$	19
Iron chelates		5-15

Reprinted in part by permission from *The Fertilizer Handbook,* 1972, a publication of the Fertilizer Institute, Washington, D.C., p. 78.

MANGANESE: Deficiencies of this nutrient have been reported in at least twenty-five states. Deficiencies of the element are most frequent in soils east of the Mississippi River where pH levels begin to exceed 6.0. The availability of this element is more strongly influenced by soil pH than either copper, zinc, or iron. If present to any extent in soil, it becomes extremely soluble and thus available. Actually it can become toxic to plants at a pH of 5.0 or less, especially if the soil contains high levels of manganese-bearing minerals. When toxic levels occur, oftentimes proper liming will reduce the toxicity hazard.

Manganese seems to be important in chlorophyll production and carbohydrate and nitrogen metabolism. The most obvious deficiency symptom is that younger leaves near the top of the plant develop an interveinal chlorosis (veins remain green but areas between are yellow). This is especially noticeable on soybeans and peanuts, and to a lesser extent on corn.

If a manganese deficiency occurs, early correction gives the greatest yield response. A soil application either with or without a mixed fertilizer before planting is probably best, as this tends to reduce the likelihood of it occurring in young plants. Usually a foliar application will give a quicker response to growing crops than that from a soil application. It has been noted on some crops, especially peanuts and soybeans, that early season leaf chlorosis may be alleviated later in the growing season apparently because of deeper root development into a more acid subsoil where more manganese may be absorbed.

Toxicity of manganese may become severe in some soils. A condition referred to as crinkle leaf in cotton is caused by manganese toxicity. The cotton leaves crinkle or pucker at the edges. Such leaves are thicker and more brittle than normal. Toxicity from this element has also been reported on tobacco and soybeans. Three available sources of manganese are given in Table 12-8.

Table 12-8
Sources of Manganese

Material		*Percent Mn*
Manganese chelate	–	12
Manganese sulfate	$MnSO_4.3H_2O$	26-28
Manganese oxide	MnO	41-68

Reprinted in part by permission from *The Fertilizer Handbook,* 1972, a publication of the Fertilizer Institute, Washington, D.C., p. 79.

The inorganic sources of manganese shown in Table 12-8 are usually equal in effectiveness on the basis of manganese content when applied to the soil. However, manganese oxide is only slightly water soluble and is not as suitable as sulfate or other highly water soluble forms for foliar application.

MOLYBDENUM: This element has been recognized only for a comparatively short time. In contrast to the other micronutrients, it is fixed in acid soils and, therefore, liming tends to make it available. Deficiencies have been described in at least twenty states.

Molybdenum is essential for fixing nitrogen by bacteria in legume nodules and in plant nitrogen metabolism. The typical symptom of yellow leaves is similar to that of nitrogen deficiency. In fact, molybdenum-deficient soybeans and peanuts (most often found on quite acid soils) are nitrogen deficient. Molybdenum deficiency is found frequently in cauliflower and broccoli. It is called whiptail and appears as irregular, long, narrow leaves. In citrus, especially trees grafted on grapefruit stock, a deficiency appears in the spring as yellow spots on leaves. If severe, these leaves die, causing defoliation.

A molybdenum deficiency may be corrected by liming soils properly. If not practical, soil additions may be made by adding sodium molybdate or molybdenum trioxide to mixed fertilizer. Recommended rates are 2 to 8 ounces of these materials per acre. For foliar applications the same amount of a molybdenum salt should be applied per acre. Also, commercially available planter-box seed treatment materials are available that are quite practical to use. For example, one ounce of a commercial source of molybdenum, such as sodium molybdate or molybdic oxide, per bushel of soybean seed is satisfactory. Sources of molybdenum are given in Table 12-9.

Table 12-9
Sources of Molybdenum

Material		*Percent Mo*
Ammonium molybdate	$(NH_4)_6Mo_7O_{24}.2H_2O$	54
Molybdenum trioxide	MoO_3	66
Sodium molybdate	$NaMoO_4.2H_2O$	39

Reprinted in part by permission from *The Fertilizer Handbook,* 1972, a publication of the Fertilizer Institute, Washington, D.C., p. 80.

ZINC: Deficiencies are quite widespread across the United States and noted on many different crops. Deficiencies have been reported in at least thirty states, with at least twenty reporting zinc deficiency for corn. Apparently this element is used in plant enzyme systems and seems to be essential for promoting certain metabolic reactions within the plant. Since zinc is not translocated from old to new tissue, deficiency symptoms first occur on the younger portions of the plant. A deficiency in corn is characterized by a white or light yellow coloration of the young bud in the early stages.

Zinc availability is reduced when soil pH is increased. Sometimes zinc deficiency is associated with high soil phosphate levels. Sometimes low temperatures and cloudy days aggravate a zinc deficiency. Its availability to plants is more important than the total amount in the soil. Like the other trace elements, zinc is required in an extremely small amount. It may be toxic to plants if excessive amounts accumulate, for example, from heavy use of some types of industrial sludge, ashes from burned buildings which had roofs or sides covered with zinc-coated metal, or from concentrated deposits of animal feces where animals have been confined to a small area and fed rations containing zinc. Zinc sources are given in Table 12-10.

Early soil and foliage applications of zinc can give large yield increases where deficiencies occur. For crops where spray programs are used (fruit and pecan trees, citrus, and rice), foliar sprays may be most convenient.

Table 12-10
Sources of Zinc

Material		*Percent Zn*
Zinc carbonate	$ZnCO_3$	52-56
Zinc chelate		9-14
Zinc oxide	ZnO	78-80
Zinc sulfate	$ZnSO_4.H_2O$	36

Reprinted in part by permission from *The Fertilizer Handbook,* 1972, a publication of the Fertilizer Institute, Washington, D.C., p. 82.

SUMMARY

This chapter was presented to give the functions in plants, forms, and common sources of secondary macronutrients as well as micronutrients. All of these nutrients are essential for plant growth and are of prime importance for maximum crop yields. Because they are often used in small quantities, they can become very toxic when applied in large amounts. Further, the balance or ratio of these materials is quite important.

REVIEW QUESTIONS

1. What functions do sulfur, magnesium, and calcium play in plant growth? What forms of these nutrients are taken up by plants? What are the common sources of these nutrients?

2. Is there any relationship between magnesium, calcium, and potassium levels in the soil? Explain.

3. What is the function of each of the micronutrients?

4. Describe the symptoms of deficiency for the micronutrients.

5. Discuss the techniques available for micronutrient application.

6. Which micronutrient(s) is(are) pH dependent for availability?

REFERENCES

Baird, J. V., Cox, F.R., and Eaddy, D. W. *Micronutrients for North Carolina Crops*. Extension Circular No. 553. North Carolina Extension Service, 1974.

Brady, Nyle C. *The Nature and Properties of Soils*. 8th ed. New York: Macmillan Co., 1974, pp. 444-55, 484-99.

Donahue, Roy L., Shickluna, John C., and Robertson, Lynn S. *Soils: an Introduction to Soils and Plant Growth*. 3rd ed. Englewood Cliffs, N.J.: Prentice-Hall, 1971, pp. 230-32, 306-07.

McVickar, Malcolm H., *Using Commercial Fertilizers: Commercial Fertilizers and Crop Production*. 3rd ed. Danville, Ill.: Interstate Printers and Publishers, 1970, pp. 87-100.

Tisdale, S. L., and Nelson, W. L. *Soil Fertility and Fertilizers*. 3rd ed. New York: Macmillan Co., 1975, pp. 261-65, 278-301, 301-41.

White, W. C., and Collins, D. C. *The Fertilizer Handbook*. Washington: The Fertilizer Institute, 1972, pp. 65-84.

13

Fertilizers and Their Application

"How important are fertilizers?"

"Over $2 billion is spent annually on fertilizers."

"You should apply them in the proper forms and amounts."

"And don't forget the importance of application methods."

In the United States, more than $2 billion is spent for plant nutrients yearly. The bringing together of nutrients for the manufacture of fertilizers is a relatively old industry; it probably began in the Baltimore, Maryland, area about 1850. Much of the impetus for development of new processes and training of personnel for the operation of processing facilities throughout the industry has been through efforts of the National Fertilizer Development Center, a component of the Tennessee Valley Authority located at Muscle Shoals, Alabama.

The scope and age of the industry should help to stress the importance of fertilizers in food and fiber production in the United States, as well as throughout the world. This chapter is designed to give the student a greater understanding of the use and value of fertilizers.

FERTILIZER TERMINOLOGY

The following terms and their meaning are quite standard throughout the fertilizer industry. As a student you should become acquainted with them.

1. Fertilizer–Any substance that is added to the soil (or sprayed on plants) to supply those chemical elements required for achieving plant growth.

2. Fertilizer material–A carrier (material) that contains at least one plant nutrient element, e.g., potassium chloride (KCl) is a potassium carrier.

3. Mixed fertilizer–A fertilizer that contains two or more of the three macronutrients, N, P or K. "Complete" mixed fertilizer contains all three macronutrients.

4. Fertilizer grade–The minimum guarantee of plant nutrient contents in fertilizer in terms of total N, available P_2O_5, and available K_2O in that order, e.g., 10-20-20.

5. Fertilizer ratio–The relative amounts of N, P_2O_5, and K_2O in fertilizers. The following examples may clarify the difference between a grade and ratio.

Grade	*Ratio*
5-10-10	1-2-2
4-8-12	1-2-3
3-9-9	1-3-3
5-15-30	1-3-6

6. Fertilizer carrier–A compound in which nutrients are contained, e.g., KCl is the form of potassium in approximately 80 percent of all fertilizers.

7. Filler–A material added to mixed fertilizers to complete the weight requirement for 2,000 pounds (1 ton).

8. Unit of plant food–Twenty pounds of a plant nutrient. The plant food unit is officially described at 1 percent of a ton.

NITROGEN SOURCES FOR FERTILIZER

In Chapter 9, it was stated that both the ammonium (NH_4^-) and nitrate (NO_3^-) forms of nitrogen were present in the soil. It was further indicated that even though both forms of nitrogen could be taken up by plants, the nitrate form was most readily utilized. Because the microorganisms in the soil usually convert

NH_4^+ to NO_3^-, it generally does not matter which form of nitrogen is applied to many crops. However the following two exceptions should be noted:

Specific Forms of Nitrogen for Flue-Cured Tobacco and Rice

FLUE-CURED TOBACCO: This crop cannot tolerate high ammonium (NH_4^+) levels from either soils or fertilizers and still produce an acceptable leaf. For this reason, it is recommended that nitrate nitrogen be used in the plant beds, and preplant fertilizers should contain at least 50 percent NO_3^- nitrogen. Side dressed nitrogen (lay by application) should be all NO_3^- nitrogen. Because the microorganisms responsible for the ammonium conversion are inefficient under cool temperatures or are practically absent in fumigated plant beds, a response to nitrogen can only be obtained by a fertilizer with a high percentage of nitrate nitrogen.

RICE: Quite in contrast to flue-cured tobacco, it is more desirable to fertilize this crop with ammonia (NH_3) or ammonium (NH_4^+) forms of nitrogen. The grower, when he places the fertilizer on the crop, has to take into account the benefits derived from keeping the nitrogen in the ammonium form until absorbed by the rice. If fertilizer nitrate nitrogen is added to the flooded soil, the reducing zone (developed from prolonged flooding) may reduce the nitrate to elemental nitrogen (N_2), and it is lost to the atmosphere as a gas.

Forms of Nitrogen for Cool Season Crops

When applying nitrogen to cool season crops such as winter wheat and spring-planted oats, it is wise, if it is economically feasible, to use a fertilizer relatively high in nitrate. This is because nitrification (conversion of NH_4^+ to NO_3^-) is reduced during the cool season due to reduced microbial activity. If ammonium is applied, the plants will take it up and convert it themselves. This conversion of NH_4^+ to NO_3^- within the plant causes a depletion of the plant's carbohydrate reserves. If much carbohydrate depletion takes place the plant can become quite yellow and stunted. Also, yellowing may be a symptom of ammonia toxicity.

Forms of Nitrogen for Lawns and Turf Grasses

Often slow-release nitrogen fertilizers are desired for lawns and turf. These nitrogen forms are desirable because they encourage less burning, streaking and last over a longer period of time. The sources of nitrogen in these fertilizers are various synthetic organic materials, such as tankage and cottonseed meal, or a plant or animal source called *urea-formaldehyde*. The desirability of these products must be weighed against their costs since they may be quite expensive.

Forms for Other Important Crops

For other crops not discussed in the previous special situations, the type will be determined largely by cost of nutrient and to some extent convenience in use. To calculate the cost per pound of N, use the following equation:

$$\text{Price/lb. of N} = \frac{\text{Price/ton of material}}{2{,}000 \text{ lb.} \times \% \text{N in material}}$$

Example: a ton of ammonium nitrate (33% nitrogen) cost $140.

$$\text{Price/lb. of N} = \frac{140}{2{,}000 \text{ lb.} \times .33} = \$0.22/\text{lb. of N}$$

The above equation is also useful when comparing the costs of various forms of nitrate nitrogen for tobacco, ammonium nitrogen for rice, and slow-release materials for turf.

Acid-Forming Fertilizers

Although to a lesser extent than formerly, some fertilizers may still be advertised as "nonacid forming." This simply means that they do not contain ammonium (NH_4^+) nitrogen or that limestone has been added to neutralize the hydrogen released during nitrification of any ammonium form of nitrogen. Acid-forming fertilizers are not a problem as long as the grower, greenskeeper, or home gardener watches the soil pH with a regular soil testing program. The amount of hydrogen released is small enough that it is usually less expensive to correct it with supplemental liming rather than purchase nonacid-forming fertilizer.

Common Sources of Nitrogen Fertilizers

Most of the usual nitrogen fertilizers found on the market today are found in Table 13-1. Again it should be stressed that for most crops cost per pound of N applied to the crop is a most important consideration.

Table 13-1
Common Nitrogen Carriers

Fertilizer	*Percent N*	*Percent NH_4 (Ammonium)*	*Percent NO_3^- (Nitrate)*	*Comments*
Ammonium nitrate	33.5	16.7	16.7	Good general-purpose material—can be used on all crops including preplant tobacco.
Ammonium sulfate	21	21	0	Use on most warm season crops if cost is low enough. Very stable in storage—provides sulfur.
Anhydrous ammonia	82	82	0	See Note 1.
Ammonium nitrate lime	20.6	10.3	10.3	Ammonium nitrate with lime added to reduce acid-forming qualities. Can be expensive. Not readily available.
Nitrate of soda	16	0	16	Common nitrogen source for tobacco.
Ammonium nitrate in aqueous ammonia	37	25	12	Anhydrous ammonia diluted with ammonium nitrate solution; stored under low pressure and gives fewer application problems than anhydrous.
Ammonium nitrate solution	21	10.5	10.5	Same properties as ammonium nitrate.
Urea	45	45	0	See Note 2.
30% N solutions	30	22.5	7.5	Very common; moderately cheap. See Note 3.
Urea formaldehyde	36.8	36.8	0	Common in lawn and turf blends. Controlled released pattern.

1. Anhydrous ammonia is a very economical source of N for corn and cotton. It is a gas under atmospheric conditions and a liquid under high pressure; thus, it requires special equipment to "knife" it into the soil. It must be placed into the soil 6 to 8 in. deep and the knife slit sealed. If the soil is quite sandy and has little CEC, losses can be expected. This material is relatively safe during application; however, it can be dangerous if poor equipment is used or careless handling occurs.

2. Urea compounds are satisfactory ammonium sources of nitrogen when placed on bare soil or tilled into the soil. If they are applied to dead vegetation, they can undergo various reactions and N be lost to the atmosphere. Urea will probably replace more and more ammonium nitrate.

3. Thirty percent nitrogen solutions are formed by mixing 50% ammonium nitrate and 50% urea. A portion of the urea of these solutions may easily be lost if they are applied to vegetation.

PHOSPHATE SOURCES FOR FERTILIZERS

Most of the phosphate used in fertilizer comes from deposits of the mineral apatite. Until relatively recently much of the phosphate used by the fertilizer industry and subsequently by producers came from apatite deposits in Florida, Tennessee, and Idaho. However, the development of the apatite deposits in Beaufort County, North Carolina, in the early 1960s provided another important source for use throughout North America, parts of Europe, and elsewhere in the world.

Phosphate fertilizers are manufactured by treating apatite (acidulating) with sulfuric, phosphoric, and nitric acid. The nature of the reactions carried out in the process are beyond the scope of this text. However, by varying certain conditions of the process, either superphosphate (20 percent P_2O_5), concentrated superphosphate (46 percent P_2O_5) or nitric phosphate (variable P_2O_5) can be obtained. The properties of rock phosphate and acidulated phosphates are as follows:

ROCK PHOSPHATE: This material consists of finely ground apatite. It has about 1 to 2 percent available P_2O_5 and cannot be used profitably except on acid soils that contain moderate to large amounts of organic matter or are highly acid, e.g., blueberry soils.

ORDINARY SUPERPHOSPHATE: This phosphate form contains 17 to 20 percent P_2O_5 and is usually sold as 0-20-0. For many years it has been the most common phosphate fertilizer but is giving way to higher analysis dry and liquid products. It contains approximately one-half gypsum and is an important source of sulfur.

CONCENTRATED SUPERPHOSPHATE: Usually this material contains 45 to 54 percent available phosphate and is commonly sold as 0-46-0. It can be used anywhere phosphate is needed; however, since it does not contain gypsum, it doesn't contribute sulfur.

NITRIC PHOSPHATE: This product (relatively new in the United States) is formed from treating rock phosphate with nitric acid. These materials were first produced at Ahoskie, North Carolina in the mid 1960s. It is a very important source of phosphorus in parts of Europe. The process in North Carolina has been discontinued because the manufacturer faced costly outlays for additional equipment to meet environmental standards.

POTASSIUM FERTILIZERS

Nearly all potassium is mined as sylvinite ore (mixture of potassium and sodium chloride) from deposits in New Mexico, Canada, Utah, and California. Potassium chloride is separated from the ore. The common name for potassium chloride is muriate of potash. It is the most common source of potassium in most mixed fertilizers. Because excessive chloride is detrimental to the quality of flue-cured tobacco leaf, a limited amount of potassium chloride can be used in mixing tobacco fertilizers. The following potassium materials are the most common ones found in fertilizers.

1. **POTASSIUM CHLORIDE**: Commonly this material is called *muriate of potash*. It is used on all crops except those sensitive to high levels of chlorine such as flue-cured tobacco. Usually it is the cheapest available source of potassium. It contains approximately 60 percent available K_2O.

2. **POTASSIUM SULFATE**: The common name for this material is *sulfate of potash*. It is a compound of potassium and sulfate (K_2SO_4) and is used on crops that are sensitive to excessive chlorine. It contains about 48-50 percent available K_2O and is used mostly on flue-cured tobacco.

3. **POTASSIUM NITRATE**: Commonly this material is known as 13-0-44. It is used mostly on flue-cured tobacco, vegetables, and citrus. It contains useful nitrate nitrogen as well as being a potassium source that does not contain chlorine.

MISCELLANEOUS FERTILIZER MATERIALS

There are numerous fertilizer materials which contain two or more elements and are frequently used by manufacturers and farmers for direct application to crops, turf, and greenhouse plants. For simplicity these materials will be listed along with their properties.

1. **Monoammonium Phosphate**: This material is completely water soluble and has an analysis of 11-48-0. It is most commonly used as a phosphate source in high-analysis mixed fertilizers and dry blends.

2. **Diammonium Phosphate**: This is another completely water-soluble phosphate material. It contains 18 to 21 percent N

and 44 to 47 percent phosphate. The most common analysis is 18-46-0. Mostly it is used in compounding high analysis and dry-blended fertilizers. However, recent studies have shown it gives a "pop-up" effect when small amounts are placed close to or with germinating seeds.

3. **Sulfate of Potash Magnesia**: This fertilizer is a mixture of potassium and magnesium sulfate. It provides about 22 percent K_2O and 18 percent MgO and has a maximum of 2.5 percent chlorine. On special crops it is a good supply of sulfur, and soluble magnesium that regular mixed fertilizers are unable to supply.

4. **Sodium Potassium Nitrate**: This is a naturally occurring material which usually analyzes about 15-0-14. It is quite useful on flue-cured tobacco, vegetables, and citrus.

FORMS OF FERTILIZER AVAILABLE TO PRODUCERS

Today's farmer is faced with an array of forms and kinds of fertilizers for providing supplemental nutrients for his crops. The following discussion will help to better understand what the fertilizer industry offers and why certain products are available. Generally speaking, the fertilizer industry produces four basic types of materials.

Pulverized Fertilizers

These are various materials containing nitrogen, phosphorus, and potassium that are mixed in varying proportions to produce the desired analysis. That is, the manufacturer will mix these carriers and add some filler (probably limestone) to produce a selected grade, such as 10-10-10 or 5-15-30. In mixing the various materials, certain chemical reactions occur. Many manufacturers add ammonia-containing nitrogen solutions or anhydrous ammonia to provide nitrogen and improve the storage and handling qualities. A conditioner may also be added to avoid caking while in storage. These fertilizers, in contrast to granulated types, are not made into uniform-sized dust-free granules.

Granulated Mixed Fertilizers

One of the main disadvantages of pulverized fertilizers is poor physical condition. If pulverized fertilizers are not properly cured, they may set up and cake when bagged. Such fertilizers are difficult to apply with conventional application equipment. They are also quite dusty. To solve some of these problems, the industry developed a granulation process. A semiliquid mixture of materials is formed. This mixture is rotated in a drum until granules form. It is then discharged into a dryer where most of the moisture is removed and the granules harden. It is cooled and screened. The finer particles are returned to the granulator; the oversized particles are crushed and returned to the screens. Currently, most of the mixed fertilizer that is not a dry blend of materials or produced as some form of fluid will be granulated. It has good physical conditions, is a homogeneous mixture with uniform-sized particles, and can be relatively high in analysis.

Dry Blends

This is simply the process of physically mixing at least two major nutrients to produce a desired analysis. This type of fertilizer can generally be produced less expensively than other types. Blenders that mix dry materials are usually located at or near the market area. Therefore, the vendor (retail sales outlet manager) frequently provides bulk handling and spreading services. Also, the blender can do a great deal of custom blending—that is, mix fertilizers to closely provide a farmer's specific need. Probably the more serious problem of dry blenders is that of separation (segregation) of the particles. Much care has to be given to proper storage, handling, transporting, and spreading of dry blends to insure that the intended grade is delivered onto the field.

Fluid Fertilizers

Liquid sources of nitrogen and phosphorus are mixed with a water-soluble solid source of potassium to produce the desired analyses. Fluid fertilizers provide homogeneous mixtures that are dust free and have excellent physical properties that make handling, storage, and spreading easy because they can be pumped around without much physical effort. Because most clear-liquid mixed fertilizers are relatively low analysis, leaders in this segment of the industry developed *suspensions* as a fluid fertilizer. Attapulgite clay or some similar material is mixed in the solution to suspend undissolved fertilizer salts, thereby increasing the analysis and more nearly competing with dry manufacturers. About the only disadvantage of this method is the additional attention that is needed to keep the salt crystals small and prevent them from growing so large in storage that they clog sprayer nozzles in application.

In order to understand how to convert pounds of plant

nutrients in fluids to a comparable amount in dry fertilizer, use a rule of thumb that the specific gravity of a liquid fertilizer is such that you can assume 11 pounds per gallon, i.e., 1 gallon of fluid fertilizer weighs about 11 pounds. For example, 9 gallons of a 5-10-10 solution will have the same nutrients in it that 99 to 100 pounds of 5-10-10 dry fertilizer will have.

Foliar Sprays or Nutrient Sprays Applied to Crop Folliage

It is not possible to apply through the foliage all the necessary N, P, and K to produce sustained high yields of most economic crops. This concept has been perpetuated by some opportunists, but sustained production is poorly documented. On the other hand, this procedure offers some hope to augment or reinforce nutrient levels of plants that are growing on soils containing incorporated nutrients. Furthermore, foliar applications do work quite well for many micronutrients such as copper, zinc, manganese, iron, and molybdenum.

FERTILIZER PLACEMENT

The correct placement or point of fertilizer application must provide the crops with an adequate supply of nutrients and still keep seedling and plant injury at a minimum. The best method is determined by several factors, such as nature of crop to be grown, type of soil, soil fertility level, and the farmer's preference. Generally fertilizer may be banded, broadcast, placed in the seed row, and/or topdressed or sidedressed. One must always be aware of the possibility of injury to germinating seedlings from either too much fertilizer or fertilizer applied in the wrong place with respect to the seed's position in the soil. Nitrogen and potassium salts in fertilizers are mainly responsible for seedling injury. Nitrogen and potassium are also strongly influenced by soil water movement. This movement is largely vertical. If the soil is wet, the movement of water and nutrients will be mostly downward. As the soil dries, water moves upward by capillary action. A rain immediately after planting, followed by a long dry period, promotes fertilizer injury. The dissolved salts move upward to a position near the roots of the germinating seedlings. If the concentration of dissolved fertilizer salts is high enough, water will be pulled out of the plant roots by osmosis. This causes a firing at the tips of the plant leaves. The plant may die if the damage is severe enough and if the weather remains dry. Death is due to a loss of cell water. The plant dries out just as if it had been placed in a heated oven. Because there are several factors to consider when selecting broadcasting and/or banding fertilizer techniques, the following discussions are presented on the two techniques.

Broadcasting

This term is used to describe a uniform application of any fertilizer or nutrient on the soil surface. The nutrient may or may not be soil incorporated. Both dry and fluid fertilizer may be spread this way. There are some distinct advantages of this method, namely that (a) large amounts of fertilizer can be applied without seedling injury, (b) labor and time are saved especially if the retail outlet applies the fertilizer, and (c) planting operations are speeded up because time spent in tending to band placement equipment at planting is eliminated.

Broadcasting fertilizer is not without some shortcomings. Broadcast phosphorus, whether a component of mixed fertilizer or a single element material, may be fixed to a high degree by certain soils. This would be more likely to occur if the soil was low in phosphorus and relatively high in clay content. Also, early crop growth may be slower since there would not be a concentration of phosphorus near the young roots. Broadcast fertilizer may also increase weeds in row middles. Modern, effective chemical herbicides will tend to eliminate this hazard.

Banding

This method of fertilizer application is that of banding or concentrating fertilizer near the seed or small transplant. An advantage of this method is that it may encourage a more vigorous, uniform, early grow-off of small seedlings. This method may cut down on the possible fixation of the phosphorus. One must be aware that band placement may cause seedling injury if too much fertilizer is used, or it is placed too close to the plant. For most crops a band that is 2 to 3 inches to the side and 2 to 3 inches below the seed is quite safe. Banding of fertilizer at planting generally slows down the planting or transplanting speed.

A slight modification of banding is to place a small amount of mixed fertilizer right in the seed row to speed up emergence. This effect will seldom give greater yields but may stimulate early growth. This practice has sometimes been termed a "pop-up" fertilizer application.

To reduce the likelihood of fertilizer band injury, always check the position of the band when planting is started. Also, the sum of nitrogen (N) and/or potash (K_2O) in the band should be noted. This is known as the amount of salt. This amount of salt should not exceed about 65-70 pounds per acre on sandy soils with not more than about 90-95 pounds per acre on clayey soils. If the amount you plan to apply exceeds this, broadcast some of the fertilizer.

Broadcasting Versus Banding

Generally where soil phosphorus and potassium levels are medium to high there is no advantage of one method over another with respect to yield increases. Even on high-fertility soils one may still expect to see some early season advantage from a band. On low-fertility soils and on heavier, clay-type soils, it may be advisable to band a portion of the fertilizer to assure an adequate supply for early growth. Some plants, like tobacco and possibly other transplanted vegetables, short-lived annual vegetables, and winter small grains may benefit from properly placed banded fertilizer.

SUMMARY

Yearly more than $2 billion are spent on fertilizers in the United States. For many crops these nutrients form a substantial portion of the production costs. Thus, it is quite important that the crop production manager have an understanding of fertilizers and the terminology used to describe them.

With the exception of rice, flue-cured tobacco, and cool-season crops, there are several sources of nitrogen fertilizer available. The exact source selected will depend on available application equipment and actual nitrogen cost. Ammonium and ammonia sources release hydrogen during nitrification. This H^+ release is termed *acid forming* and increases soil acidity. However, there isn't sufficient soil acidity to warrant using more expensive nitrate sources of nitrogen.

There are four sources of phosphate available. These include rock, ordinary super, concentrated super, and nitric phosphates. Rock phosphate is low in available P_2O_5, except on highly acid soils, and nitric phosphates are rather new in the United States. Either ordinary or concentrated superphosphates are quite suitable phosphate materials.

The most common potassium fertilizer material is muriate of potash. It is quite suitable for all crops except those sensitive to chlorine. If chlorine is a problem, potassium sulfate or potassium nitrate are very acceptable potassium sources.

In addition to the major N, P, and K sources of fertilizers, monoammonium phosphate, diammonium phosphate, and sulfate of potash magnesia are miscellaneous fertilizer materials. These materials may be used alone or to blend mixed fertilizers.

Crop managers have available to them pulverized mixed fertilizer materials, granulated mixed materials, dry-blended materials, fluid (liquid) fertilizers, and foliar sprays. With the exception of foliar sprays, all sources are equal in effectiveness, provided they are applied properly. Usually, sufficient quantities of macronutrients cannot be applied by foliar sprays. Certain micronutrients can be applied by foliar sprays.

When applying fertilizers, band and/or broadcast techniques may be used. Either technique is quite successful for small amounts of nutrients. However, fertilizer levels containing over 65 to 70 pounds of salt should probably be broadcast on sandy soils and if the salt levels reach 90 to 95 pounds, it would be well to broadcast on clay soils.

REVIEW QUESTIONS

1. Define: fertilizer, fertilizer material, mixed fertilizer, fertilizer grade, fertilizer ratio, fertilizer carrier, filler, and unit of plant food.

2. What forms of nitrogen should be applied to flue-cured tobacco? Rice? Why?

3. How does ammonium nitrogen affect cool-season crops?

4. What form of nitrogen is most suitable for turf grasses? Why?

5. How can you best determine the cheapest form of nitrogen for most crops?

6. What are acid-forming fertilizers?

7. Give the percent and forms of nitrogen in the following material: ammonium nitrate, ammonium sulfate, anhydrous ammonia, nitrate of soda, urea, ammonium nitrate solution, and urea formaldehyde. Should special precautions be taken when using these materials?

8. How are most 30 percent nitrogen solutions formed?

9. What are the percentages of available P_2O_5 in rock, ordinary super, concentrated super, and nitric phosphates?

10. What are the percentages of available K_2O in potassium chloride, potassium sulfate, and potassium nitrate?

11. Can potassium chloride be used on all crops?

12. What fertilizer form is used as a "pop-up"?

13. Contrast the use of pulverized, dry blends, granulated mixed, and fluid fertilizers.

14. When should you use foliar sprays?

15. Define: fertilizer banding and broadcasting. What are the advantages of each?

16. How much fertilizer should be placed in a band?

REFERENCES

Brady, Nyle C. *The Nature and Properties of Soils*. 8th ed. New York: Macmillan Co., 1974, pp. 504-20.

Donahue, Roy L., Shickluna, John C., and Robertson, Lynn S. *Soils: an Introduction to Soils and Plant Growth*. 3rd ed. Englewood Cliffs, N.J.: Prentice-Hall, 1971, pp. 297-328.

Foth, H. D., and Turk, L. M. *Fundamentals of Soil Science*. 5th ed. New York: John Wiley, 1972, pp. 299-325.

McVickar, Malcolm H. *Using Commercial Fertilizers: Commercial Fertilizers and Crop Production*. 3rd ed. Danville, Ill.: Interstate Printers and Publishers, 1970, pp. 35-77.

Olson, R. A., et al., *Fertilizer Technology and Use*. 2nd ed. Madison: Soil Science Society of America, 1971, 221-37, 273-300, 303-32, 338-56, 381-400, 414-45, and 456-83.

Tisdale, S. L., and Nelson, W. L. *Soil Fertility and Fertilizers*. 3rd ed. New York: Macmillan Co., 1975, pp. 365-85, and 505-22.

14

Soil Classification

> *"What* is *soil classification?"*
>
> *"A systematic grouping of soils into classes, based on soils' properties."*
>
> *"It organizes our knowledge."*
>
> *"It helps us remember soils' properties."*

Whenever information is obtained or data collected on any item, this information quickly becomes voluminous and hard to comprehend. When this happens a classification or grouping system is needed. For example, suppose soils were never grouped but a hole was bored in the landscape and some information about this very small amount of soil was collected. As the number of borings became large, it would be necessary to group certain borings into classes with certain similar characteristics. By doing this we could remember the properties of the groups rather than the properties of each individual. Thus, it can easily be seen that by classifying soils we can do the following:

1. Organize our knowledge relative to groups of pedons.
2. More easily remember soil properties associated with different soil series.
3. More easily study and describe similarities and differences between various pedons.

SOIL CLASSIFICATION SYSTEMS

The roots of modern soil classification are usually attributed to a Russian soil scientist, V.V. Dokuchaiev. During the late 1800s, Dokuchaiev developed the concept that soils are natural bodies formed by the factors of soil formation. Because of Dokuchaiev's early work, many of the soil classification systems throughout the world developed around soil as a natural body formed by the processes of nature. These soil classification systems recognized processes of soil formation and classified soils according to the evidence of these processes.

The United States has been through three systems of soil classification. Originally the U.S. Department of Agriculture in cooperation with the land grant universities in each state classified the soils in each state to inventory the nation's resources. The classification system used was of a descriptive nature and consisted of identifying soils in terms such as brown, loamy forest soils or black prairie soils. Later the soil classification system was refined and soils given names of towns and landmarks where they were first found. This classification developed until 1938 when it was replaced by a more detailed system developed by the U.S. Department of Agriculture.[1] Both the earlier system and the 1938 system allowed soil formation processes to dominate the classification.

By 1960, the 1938 system had become limiting due to new knowledge in an ever-expanding field of study. Thus, a new system was tentatively adopted. The 1960 system recognizes the processes of soil formation but actually classifies soils according to their properties.[2] In the years following 1960, the classification system was tested and expanded until it was published in a somewhat final form in December 1975.[3]

This modern system consists of six levels of classification which starts with very broad groups of soils and moves downward to the individual soil with a distinct set of properties. Briefly, this system is as follows:

- Orders
  - Suborders
    - Great Groups
      - Subgroups
        - Families
          - Series

---

[1] M. Baldwin, C. E. Kellogg, and J. Thorp, "Soil Classification," in *Soils and Men*, 1938 Yearbook of Agriculture (Washington, U.S. Dept. of Agriculture, U.S. Govt. Printing Office, 1938), pp. 979-1001.

[2] Soil Survey Staff, *Soil Classification, a Comprehensive System—7th Approximation* (Washington: U.S. Department of Agriculture, U.S. Govt. Printing Office, 1960).

[3] *Soil Taxonomy: A Basic System of Soil Classification for Making and Interpreting Soil Surveys*, Agriculture Handbook No. 436 (Washington, D.C.: Soil Survey Staff, Soil Conservation Service, U.S. Department of Agriculture, Dec. 1975).

There are ten soil orders in the classification system. Because the orders are the highest level of classification, each group is very broad and is different from other orders because of some major set of soil-forming processes. The ten orders and a brief description of each order is given in Table 14-1.[4]

Table 14-1
Soil Orders and Their Characteristics

Order	*Description*
Alfisols	Unleached forest soils of northern and midwestern U.S.; well-developed A2 and B horizons present; base saturation usually above 35% in the B horizon.
Aridisols	Soils of arid regions; insufficient moisture to produce enough organic matter for thick O or A1 horizon; bases are not leached and accumulate in the A horizon because evaporation is greater than leaching; common in southwestern U.S.
Entisols	Recent soils that show little evidence of the factors of soil formation; common in alluvial areas and on steep slopes.
Histosols	Organic soils; see Chapter 3 for the extent and formation of organic parent materials.
Inceptisols	Recent soils that show beginning evidence of soil formation; clay and bases have not moved to any extent; B horizons may be red colored and calcium carbonate may have leached.
Mollisols	Soils with thick A horizons; usually developed under grass vegetation; base saturation above 35%; usually thought of as the prairie soils of the Great Plains.
Oxisols	Deep red soils of the tropics; usually quite old and highly leached; high in iron.
Spodosols	Soils in which aluminum, iron, and humus have moved into the B horizon; often called *podzols* due to the influence of the 1938 classification system; scattered throughout the eastern U.S.
Ultisols	Very similar to the alfisols except base saturation is 35% or less; formed in warm, moist climates; very common throughout the southeastern U.S.
Vertisols	Soils high in 2:1 clays; the clays shrink and swell enough to mix the horizons; common in the black

[4] Ibid., Chap. 8-17.

belt soils of the Mississippi Delta and the black clays of Texas.

The names of the soil orders are formed by adding a formative element and a connecting *o* or *i* to the word *sol* (Latin for soil).

Suborders are a refinement of the orders on the basis of characteristics such as wetness and temperature. Presently forty-seven suborders are recognized in the United States.[5] The suborders are named by adding a connotative element to the formative elements of the orders (i.e., an aquult would be a wet (aqu) ultisol (*ult*-i-sol).

Great groups are subdivisions of suborders based on soil horizons and their arrangement, temperature regimes, and similar base saturations. About 185 great groups are recognized in the United States.[6] The great group name consists of adding a one to two element prefix to the suborder name. This prefix is used to suggest something about the horizonation, temperature, or base status in the great group.

Subgroups are refinements of the great groups to include the central concepts of the great groups; great groups which grade toward other orders, suborders, and great groups; and great groups of soils which fit neither in the central concept, nor grade to other areas of the classification system. There are about 970 subgroups in the United States.[7] The subgroups are named by adding the adjective typic to the great group name where the subgroup represents the central concept of the great group. When the subgroup represents a gradational form of the great group, the adjective form of a gradational group name is used to modify the name of the great group. Other adjectives may be added to the great group names when they are needed to describe a great group which fits neither the central concept nor gradational phases of the great group.[8]

The soil family is a group of soils within a subgroup. The soils are grouped into families on the basis of properties that affect soil use (usually agricultural and/or engineering). In the United States about 4,500 soil families are recognized.[9] Soil family names are derived by using descriptive terms which reveal texture, mineralogy (type of clay), soil temperature, thickness of

---

[5] Ibid., p. 77.

[6] Ibid., pp. 77-8.

[7] Ibid., p. 79.

[8] It is realized that soil classification terminology is above the scope of the text. It is simply mentioned so that the student knows it exists and can possibly know what is being discussed when soil group names are mentioned in modern literature.

[9] *Soil Taxonomy*, p. 80.

root zone, and other properties which affect plant growth and engineering properties.

Within soil families we find soil series. The soil series is basically the modern-day product of the five factors of soil formation. It is the lowest recognized category in the soil classification system. Today about 10,500 soil series have been described and recognized in the United States.[10] Soil series are named after the area, town, or community where they were first found.

The properties that determine the soil series are often the same as those that determine the higher groups. However, more properties are used and the allowable range for each series is rigidly defined. The properties that determine the soil series are as follows:

1. Horizon order and sequence
2. Horizon development or thickness
3. Texture of each horizon
4. Organic matter content
5. Soil pH of each horizon
6. Parent materials
7. Depth to hard rock
8. Pan horizons present
9. Soil color
10. Soil structure
11. Type of clay present
12. Any other factor that would make the soil different.

When using the soil classification system in land use planning, we need to select a level of detail which will fit our needs. Usually worldwide planning uses the orders and suborders; geographers often find the great groups and subgroups very helpful. Broad-scale agricultural planning finds the soil family very useful (i.e., country agricultural uses, etc.). Individual farm plans may need the soils classified at the series level while intensive agricultural and engineering uses may require classification of phases of soil series.

Two soil classes that are very common should be mentioned at this time. The soil association is a soil class which simply refers to how soils occur on the landscape. The soil association name

[10] Ibid., p. 80.

consists of the names of the soil series occurring in the soil association. To use soils classified according to soil association we must refer back to the soil series in the association.

A second common classification is the soil type. Originally the soil type was the lowest level of classification in the 1938 system.[11] The soil type was determined by the surface texture of the soil. Today the soil type is simply included as a phase of a soil series whenever soil series refinement on the basis of soil surface texture is necessary.

USING SOIL CLASSIFICATION

Soil survey is the technical application of soil classification. A soil survey is conducted by dividing a landscape into areas containing similar pedons or polypedons. Usually the landscape areas are divided on the basis of soil series, slope and erosion, and recorded on an aerial photograph. The factors which determine the soil series were discussed previously. The slope classes used may vary but usually fall in the ranges defined in Table 14-2.

Table 14-2
Soil Survey Slope Classes

Slope Class	*Percent Slope**
A	0-2
B	2-6
C	6-10
D	10-15
E	15-25
F	-25+

*The percent slope is defined as the number of feet of rise or fall in 100 feet. It may be calculated as follows:

$$\text{Percent slope} = \frac{\text{feet fall}}{\text{horizontal distance}} \times 100$$

Erosion classes are very complicated and are used only when necessary. Originally the erosion classes were based on the amount of topsoil which had been removed from the soil. This was no problem as long as virgin soils were present to determine the

[11] Ibid., p. 81.

original amount of topsoil. Today it is often very hard to tell how much topsoil has been removed. Thus, we use the working definitions of the erosion classes presented in Table 14-3.

Table 14-3
Soil Survey Erosion Classes

Erosion Class	*Official Definition*	*Working Definition*
0	Topsoil washed in	15 in. topsoil
1	¼ topsoil removed	10-15 in. topsoil
2	¼-¾ topsoil removed	3-10 in. topsoil
3	¾ topsoil removed	3 in. topsoil
4	Gullied land	Gullied land

Once the soil series, slope, and erosion have been determined, they are recorded on a soil map (aerial photographs) as a mapping unit. These mapping units consist of a number or a pair of letters for the soil series, a capital letter for the slope class, and the erosion class number. They will always occur in this fashion unless the soil series is one which has no erosion or slope class. If slope and erosion classes are absent only the soil series number or abbreviation is used. Two examples of mapping unit symbols are as follows:

Ce Dl = Cecil series, 10-15% slope, ¼ topsoil removed

and

Ap C2 = Appling series, 6-10% slope, ¼-¾ topsoil removed

A portion of a soil survey map is shown in Figure 14-1.

Once the soil survey map is prepared, it is usually combined with additional materials and called a *soil survey report.*

Soil Survey Reports

In addition to the mapping units recorded on aerial photographs, a soil survey report contains other materials including the following:

1. Features on the map such as:
   a. Cities
   b. Roads

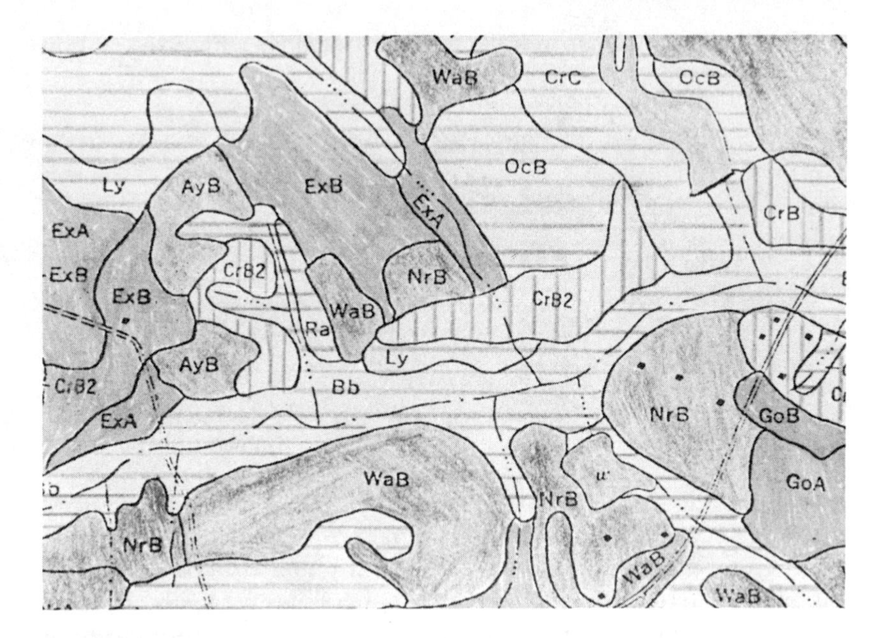

Figure 14-1. The soil survey map outlines the occurrence of mapping units on the landscape.

c. Houses

d. Streams

e. Churches

f. Other landscape features such as mines, factories, and cemetaries

2. Supplementary materials

After the soil map is made and soil samples are collected for research data, another section is added to the soil survey report. This section contains the following:

1. Descriptions of all soils

2. Guides for their use and management

Thus we see the soil survey report is readily usable as a soil inventory for land managers, appraisers, researchers, farm planners, urban developers, and soil engineers.

SOILS INFORMATION AVAILABLE TO THE LAND MANAGER

There are three general levels of information available to the land manager. The type he chooses to use will depend on the amount of detail required. In general, the following three types of information are available:

1. General state and county maps—Broad-scale association maps for planning general land use patterns
2. County soil survey maps—Detailed maps with interpretations for each mapping unit
3. Farm maps—Very detailed maps with specific soil interpretations for each soil series in each field

With intensive land use patterns developing in urban areas, it is often necessary to obtain detailed soil maps on a lot-by-lot basis. These maps are usually prepared from detailed borings and may employ phases of soil series to obtain detail.

SOURCES OF SOIL SURVEYS

Soil surveys are usually conducted by the U.S. Department of Agriculture Soil Conservation Service in cooperation with the land grant universities and the Agricultural Extension Service in each state. Although it is a cooperative effort between the three organizations, the Soil Conservation Service usually assumes the responsibility for the mapping and report preparation activities. The land grant universities provide research data and the Extension Service assists in educational endeavors.

General state and regional maps as well as county soil surveys are prepared by these three agencies. The Soil Conservation Service also prepares individual farm maps for farmers that have organized into soil and water conservation districts. On a time available basis, the Soil Conservation Service will assist urban planners and developers with soil inventories and interpretations. In some states the Agricultural Extension Service assists in soil surveys and land use interpretations for individual land managers.

Even though considerable soil survey assistance is available from the previously discussed state and federal agencies, the demand has seriously outdistanced their capabilities. To meet the increased demand for soil interpretations, many counties and municipalities have employed soil surveyors to help with soils problems in sanitation and urban development. Engineering consulting firms are adding qualified soil surveyors to their staffs so that they can offer a broader range of services for urban developers.

With the increased demand for soil surveys and soil survey information, a severe shortage of soil surveyors has developed. Thus, unqualified or inexperienced individuals have conducted soil surveys. As a safegard against poor soil interpretations and/or inferior soil surveys, many states are creating licensing boards to impose standards and license soil surveyors. These licensing agencies usually develop their technical standards along guidelines set by the Soil Conservation Service and the land grant universities.

LAND USE EVALUATION

A discussion of soil classification would not be complete without mentioning land use evaluation systems or land classification systems. In addition to classifying soils based on their properties, we can also classify them according to their potential for certain uses. Many of these systems have developed. Only the land capability system developed by the Soil Conservation Service will be discussed.

The soil series is a very useful tool in describing small areas of soil for urban development, soil engineering, and detailed agricultural production. Soil families and to some extent the soil association are useful in describing broad general soil areas for regional plans. However, a farmer may find it necessary to group several soil series simply to have fields large enough for efficient machinery operations. A land unit which groups several soil series into usable areas is the land capability class.

Land Capability Classes

Soil series, which are similar with respect to the physical, chemical, or mineralogical properties mentioned in the previous section, can be expected to have many of the same drainage and permeability properties. If these similar soil series are collected according to slope and erosion, we have a land capability class. Therefore we can say capability classes are groups of soil series which are similar with respect to slope and erosion. The Soil Conservation Service of the U.S. Department of Agriculture recognizes eight different land capability classes according to the slope the soil series has developed on. These eight classes are described in Table 14-4. This table also notes the slope class and land use for these capability classes. The conversion of any mapping unit to a land capability class is quite simple if a few rules are observed.

Table 14-4
Land Capability Classes

Class	*Percent Slope*	*Slope Class*	*Description*
I	0-2	A	Nearly level land with few limitations; row crops may be grown every year.
II	2-6	B	Gently sloping land with ¼ to ¾ of the topsoil removed; row crops every other year.
III	6-10	C	Moderately sloping land with ¼ to ¾ of the topsoil removed; row crops one out of every three years.
IV	10-15	D	Sloping land with ¼ to ¾ of the topsoil removed; row crops one out of every four years.
V	–	–	Land not suitable for agriculture because of some special hazard such as flooding or drought (could become Class I if the problem were removed).
VI	15-25	E	Very sloping land subject to severe erosion; only suitable for trees and light grazing.
VII	25	F	Mountain slopes not suitable for cropland; may be used for forestry and recreation.
VIII	–	–	Mountain peaks and coastal marshes; not suited for agricultural production; may be quite valuable for recreation.

If the erosion class of a mapping unit is two (2), the slope classes A, B, C, and D, convert directly to capability classes I, II, III, and IV, respectively. If the erosion is less than 2, then the capability is moved to a class which allows more intensive land use. If the erosion is greater than 2, then a less intensive land use capability class is required. A few examples will clarify these points.

Mapping Unit	*Capability Class*
Ce C2	III
Ce C1	II
Ce C3	IV
Ce D2	IV
Ce D3	VI (note: class V is special)
Ce A2	I
Ce A1	I

From Table 14-4 the land use for each capability class can be observed. These uses do not tell us which crops to grow. They simply tell us the intensity with which the land can be cropped and yet prevent erosion. The beginning student should not confuse the land capability class with the soil series as a tool for determining specific land uses.

SUMMARY

This chapter was presented to make the student aware of the modern soil classification system and its application to soil survey. It was also presented to discuss soil surveys and their various uses in land use planning. Whenever entering into land use planning or land management decisions, it is wise to consult your local Soil Conservation Service for available soil survey information and interpretations.

REVIEW QUESTIONS

1. What is the purpose of a soil classification system?
2. What are the six levels of the modern soil classification system?
3. What is a soil association?
4. What properties determine the soil family? The series?
5. What are soil slope classes? Can you calculate percent slope?

6. What are soil erosion classes?

7. What is a mapping unit?

8. What supplementary materials are contained in a soil survey map?

9. What could a soil survey be used for?

10. What types of soils information are available?

11. Who is responsible for soil surveys?

12. What is a land capability class?

13. Can you convert mapping units into land capability classes?

14. Contrast the uses of land capability classes, soil series, and soil families.

REFERENCES

Berger, Kermit C. *Introductory Soils*. New York: Macmillan Co., 1965, pp. 105-26.

Brady, Nyle C. *The Nature and Properties of Soils* 8th ed. New York: Macmillan Co., 1974, pp. 317-52.

Buol, S. W., Hole, F. D., and McCracken, R. J. *Soil Genesis and Classification*. Ames: Iowa State University Press, 1973, pp. 170-322.

Donahue, Roy L., Shickluna, John C., and Robertson, Lynn S. *Soils: an Introduction to Soils and Plant Growth*. 3rd ed. Englewood Cliffs, N.J.: Prentice-Hall, 1971, pp. 101-75.

Foth, H. D., and Turk, L. M. *Fundamentals of Soil Science*. 5th ed. New York: John Wiley, 1972, pp. 231-72.

Simonson, R. W., ed. *Non-Agricultural Applications of Soil Surveys. Developments in Soil Science 4*. Amsterdam: Elsevier Scientific Publishing Company, 1974.

Soil Taxonomy: A Basic System of Soil Classification for Making and Interpreting Soil Surveys. Soil Survey Staff, Soil Conservation Service, U.S. Department of Agriculture. U.S. Department of Agriculture Handbook No. 436, Dec. 1975.

Soil Survey Manual. U.S. Department of Agriculture Handbook, No. 18: Supplement. U.S. Department of Agriculture, May 1962, pp. 1-42, 409-34.

15

Soil Conservation

The treatment of soil conservation in this text will be very brief. This section is presented to make the student aware that a problem exists and that many solutions are already available. If the manager is quite serious about conservation, he can consult with his local Soil Conservation Service representative and agricultural extension agent and obtain a great amount of assistance.

Modern soil conservation is concentrated in three major areas. These are controlling soil erosion, stopping soil pollution, and regulating land use to insure that all land is used as efficiently as possible.

EROSION CONTROL

A few years ago, soil erosion was deemed a farm problem that did not concern anyone but farmers. Today, this type of thinking has nearly vanished and many people have begun to appreciate the problems of erosion in urban and recreational areas. Simply stated, erosion consists of either wind or water transporting soil from one place to another. To the farmer, this usually means losing topsoil from his productive lands and possible siltation of his farm ponds and streams. To the urban land user, it means unsightly eroded areas in urban developments, muddy streets, and siltation in sewage lines and reservoirs. To the naturalist, erosion means siltation of lakes with soils and pollutants carried in the soils and ugly scars on the landscape. With the previous comments in mind, we are now ready to look at some of the factors that cause wind and water erosion and some possible control measures.

Water Erosion

Water erosion takes place whenever flowing water passes over a loose soil and carries some of it away. There are four basic types of water erosion:

1. Splash erosion—Raindrops strike the soil surface and break soil aggregates into fine particles which can be carried away
2. Sheet erosion—Water moves across the soil surface and removes thin sheets of soil
3. Rill erosion—Water moves across the soil surface and cuts many small ditches a few inches across
4. Gully erosion—Water flows across one spot long enough to cut large gullies

Although all of these types of erosion are important, the sheet and rill types are very dangerous because they are not always obvious until all of the topsoil is gone. To better understand how all these types of water erosion take place, let us briefly review the factors that are responsible for them.

1. Amount and distribution of rainfall—The amount of erosion that will take place is determined both by the amount and the time it takes for a given rain. Large amounts of rainfall in a short period of time can cause severe erosion, but the same amount of rainfall over a long period may not cause any erosion.
2. Slope of the land—The slope of the landscape controls the velocity of runoff water. Thus, steep slopes erode more rapidly than relatively flat areas.
3. Size of the watershed—The rate at which a drainageway will erode will depend greatly on the size of the watershed behind the drainageway. Certain steep slopes

Figure 15-1. This field shows evidence of sheet, rill, and gully erosion.

can stand fairly high velocity rains provided these slopes drain only small areas. Likewise, severe erosion can occur on 2 to 3 percent slopes if they are long and drain large acreages.

4. Soil characteristics—Those soil characteristics which affect infiltration and percolation (soil texture, structure, and consistence) also control erosion. Water runoff is increased as infiltration decreases. This increase in runoff causes increased erosion. (See Chapter 4.)

5. Vegetative cover in the watershed—Even on steep slopes, heavily vegetated watersheds seldom erode. The erosion potential is greatly increased as row crops become prominent and the soils are not vegetated during some portion of the year.

All of the previous factors will determine how much erosion will take place. Thus, they should all be considered in developing an erosion control program.

Wind Erosion

Wind erosion is usually a sheet erosion that leaves behind drifts of soil and holes, or blowouts. Until a few years ago, this type of erosion was not important except in certain western states. With the incorporation of small farms into larger units and the creation of large fields, this problem is becoming important because many windbreaks have been removed. Today wind erosion is becoming a problem in both the Midwest and Southeastern Coastal Plain.

Figure 15-2. Increased field sizes have removed sufficient windbreaks to cause severe wind erosion in many parts of the United States.

Erosion Control Measures

A large amount of information is available on erosion control on different soil series, vegetative covers, and slopes. The U.S. Department of Agriculture Soil Conservation Service has a Universal Soil Loss Equation which can be employed to calculate the expected erosion over most soil conditions.[1] From this equation and other information, many erosion problems can be quite readily overcome.

The following control measures are given to help solve minor erosion problems. If erosion becomes a real problem, seek professional advice through the Soil Conservation Service.

MECHANICAL CONTROL MEASURES:

1. Establish vegetated waterways.
2. Use contour tillage.
3. Construct terraces (works on slopes up to about 12 percent).
4. Use diversion ditches and dead furrows to remove water across rather than down the slope.

VEGETATIVE CONTROL MEASURES:

1. Crop according to the intensity determined by the land capability class.
2. Use strip crops.
3. Use soil mulches and no-tillage planting techniques.
4. Increase soil structure if possible.
5. Add organic matter to the soil whenever possible.
6. Grow cover crops in the winter.
7. Plant windbreaks.
8. Plant annual grasses on construction sites that cannot be permanently seeded.

SOIL POLLUTION AND WASTE MANAGEMENT

The soil acts as a purification system that is capable of holding many pollutants and tying them up so that they are either rendered inactive or biodegraded. However, the soil's holding capacity (cation and anion exchange capacity) can quickly be saturated

[1] *The Universal Soil Loss Equation with Factor Values for North Carolina*, Technical Guide, Section II-D (Raleigh: U.S. Department of Agriculture, Soil Conservation Service, Oct. 1976).

and the soil sterilized to a barren, polluted, unproductive area which can erode and become a further source of pollution. Soil pollution is not limited to any one phase of soil use. Agriculture, industry, and urban development all contribute to soil pollution.

In agriculture, soil pollution can result from improper use of agricultural chemicals (including fertilizers) and the improper management of waste products. The previous should not be construed to mean that agricultural chemicals are posing a threat to our soil when they are handled properly. Pollution arises from improper applications due to poor equipment, pure neglect on the part of the applicator, improper chemical container disposition, and improper disposition of leftover chemicals.

Presently, very stringent federal and state laws are being instituted to deal with pollution from agricultural chemicals. These laws include educational programs, licensing of applicators, inspection of chemicals and equipment, and mandatory punishment of violators.

In addition to pollution from agricultural chemicals, agricultural pollution also arises from soil erosion (covered earlier in this chapter) and animal waste management problems. Animal manures are an excellent source of nutrients. However, concentrations of livestock in small areas can cause excessive accumulations of waste materials. If these waste materials are returned to the soil they are quite useful as fertilizers. If they are allowed to erode or move into streams they can become harmful to the environment. Today many states require adequate animal waste disposal systems before they will allow installation of such facilities.

It should be remembered that animal waste pollution is usually a concentration of useful nutrients that may be beneficial if it can be redistributed to the soil. These wastes are easier to handle than industrial, urban, and municipal wastes which may contain toxic substances harmful to the soil.

The topic of industrial waste is very complex. It ranges from waste consisting of warm water, to biological and organic materials, to toxic inorganic components. Strong social concern over the environment has demanded through legislation that these wastes be contained or neutralized by the industries responsible for them. Waste treatment facilities range from simple holding ponds for evaporation to complex chemical systems designed to remove toxic components from waste water, smoke, and sludge. Once the toxic substances are removed, the remaining water and sludge are usually returned to the soil to complete the removal of nitrogen, phosphates, and other plant nutrients. The installation and operation of these facilities are quite expensive. Many industries are faced with large capital outlays to comply with pollution laws.

Urban and municipal waste problems are involved with handling the waste products generated by population concentrations. These waste materials fall into two categories: (1) solid waste and (2) human waste.

Today most solid waste is contained in landfills. The landfill consists of a mass of compacted solid waste encased in clayey materials so that water cannot percolate into or out of the waste. The exact construction of a landfill will depend on local soil conditions. In areas of deep, well-drained soils, the waste will be placed in a trench (Figure 15-3), compacted, and covered daily with 4 to 6 inches of soil. When the trench is full, a layer of clayey

Figure 15-3. The sanitary landfill is a common method of solid waste disposal.

materials will be applied and compacted. Topsoil will then be graded over the clay and the area reseeded to grass, crops, or trees. On low, poorly drained areas the landfill is often placed on top of the soil to avoid water table contamination. In this case the completed landfill is a mound of covered waste material.

A schematic of a completed landfill is shown in Figure 15-4. A detailed knowledge of soil properties is very important in se-

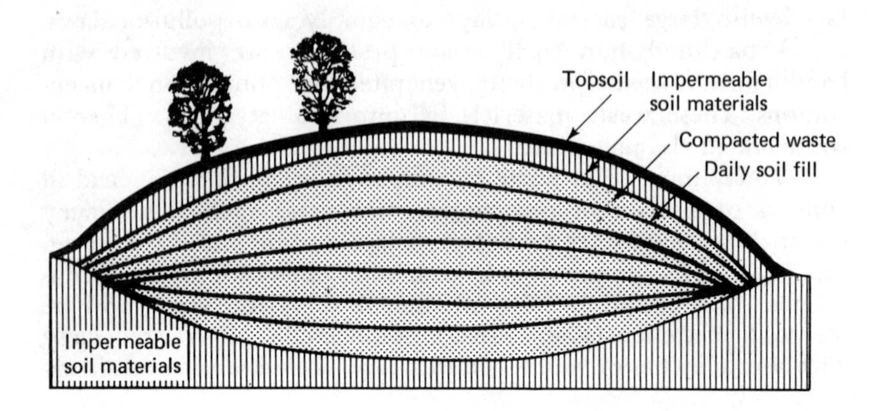

Figure 15-4. Cross-section of a completed landfill.

lecting landfill sites and determining operational specifications.

Although the landfill is a satisfactory method of waste confinement, it uses valuable land and is subject to failure due to construction techniques. It is probably a short-lived technique that will be replaced by waste management technology capable of recycling the waste products.

The disposal of human wastes is handled by the septic tank in urban areas and the municipal sewage system within towns and cities. The average septic tank receives the household wastes from the bathroom, laundry room, and kitchen sink. These wastes are held in the septic tank until anerobic microorganisms break down the organic components to solids, which settle to the bottom of the tank, and liquids which are pushed into a filter field for aerobic degradation (see Figure 15-5).

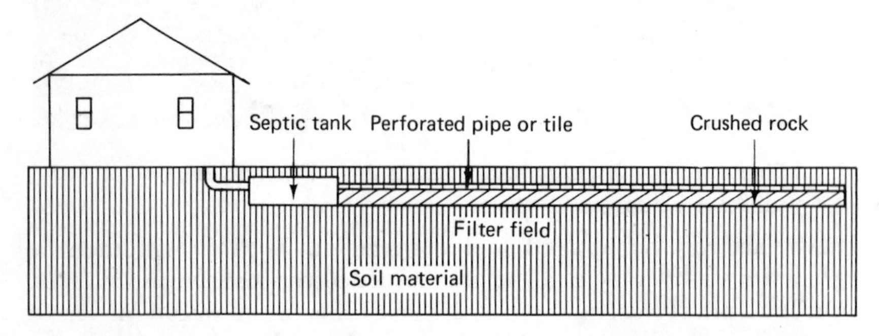

Figure 15-5. Schematic of a septic tank and filter field. The filter field may consist of several tile lines.

Soil characteristics of the filter field determine the field size and efficiency. Filter fields seldom work satisfactorily in clayey soils. The effluent (waste water) will rise to the surface and become a source of pollution. On sandy soils the effluent will percolate through the soil and may contaminate wells and ground water supplies (Figure 15-6). Local governments are usually in charge of approving septic tank sites and installations. These agencies often hire soil scientists to contribute soils information on marginal land areas for septic tanks.

Figure 15-6. Unless proper soil combinations are present, these septic lines can cause a future health hazard.

The municipal sewage treatment plant receives sewage from the same sources as the septic tank, plus it often handles waste from business concerns and small industries. The full process of the sewage disposal plant is quite complex, but generally it consists of digesting the solid materials to sludge and purifying the waste water. The sludge is then returned to the land and the waste water is chlorinated and often returned to the stream from which it was extracted. If the water purification facility does not remove all nutrients, the waste water is often irrigated on to land areas for further purification. The characteristics of the soil receiving the sludge and waste water are very important because they determine the acceptable rates of application. If these materials are overapplied, severe problems may result.

From the previous discussion it should be obvious that soils play an important role in waste disposal and management. If the role of soils in these areas is taken lightly, our soils may become severely polluted from many sources.

LAND USE REGULATIONS

With the ever increasing population growth pressures, the need to use all land very efficiently is quite important. Formerly lands such as mountain peaks, inland swamps, and tidal marshes were thought of as worthless. However, it has been shown that all of these areas greatly contribute to our ecological system. Furthermore, these areas are often very fragile and any disturbances such as cutting timber, draining swamps, and dredging marshes often temporarily or even permanently destroy them. Because of the importance of these areas, local and state governments as well as the federal government have taken steps to impose land use regulations for the use of these areas. For example, environmental impact statements must accompany all requests to dredge most areas along the Atlantic Coast.

In addition to the destruction of fragile ecological systems on some land use areas, the United States is faced with the problem of urban and industrial development competing with agriculture for prime land. Usually prime agricultural soils are never again used for food production once they have been taken for highways and/or urban and industrial uses. This competition for agricultural lands is depleting our food production base, as well as forcing marginal lands into agricultural production. Although the short-term effects of diverting prime agricultural lands to other uses may greatly help certain portions of our society, the long-term effects are higher food prices and insufficient food production.

To combat these problems of the destruction of fragile areas and a decreasing agricultural base, governments at all levels are looking at land use regulations schemes. These methods of land use regulation usually are one of the following, or a combination thereof:

1. Zoning

2. Differential taxation
3. Land use legislation

Zoning is a land use regulation tool that is most often used to regulate the use of land in and around towns and cities. Essentially, zoning ordinances state the allowable uses for certain parcels of lands. If the landowner wishes to use these lands for other purposes, he must request permission from the local government, present an explanation of these uses at a public hearing; and abide by the decision of the governing body on his request. If he feels he has been treated unfairly, his recourse is the court systems of the nation.

Differential taxation is a land use regulation tool often used by counties to keep prime lands in agriculture. It is customary that as cities expand, the land values around the cities inflate to the point that the owner must sell or subdivide to pay the property taxes. By allowing the owners of prime farm lands an agricultural tax base, growth and development is often forced to less than prime agricultural lands. Of course, most differential taxation plans contain provisions that the owner of the lands once declared agricultural must pay a penalty if they are later used for other purposes.

Land use legislation is a tool most often used by state and federal governments to protect certain fragile areas from destruction. Nearly all land use legislation is more general than zoning and aims at enforcing certain ideals rather than protecting or regulating given parcels of land. At the time of this writing, most land use legislation in the United States is poorly coordinated and doesn't follow the same patterns from state to state. However, the problems of land use regulation occupy the efforts of state governing bodies yearly, and for several years bills for federal land use regulation have been brought before Congress.

In addition to the distinct areas of zoning, differential taxation, and land use legislation, there are several miscellaneous land use regulation techniques. These techniques involve such things as health departments regulating septic tank drainage field sizes, the regulation of sanitary landfill sites, laws to require park construction by urban developers, and the providing of natural areas in subdivisions, etc. Many of these land use regulation techniques are based on the assumption that certain soils are best suited for certain uses.

All land use regulation techniques contain a common thread.

They restrict the landowner's right of land use for the benefit of society. These regulations are rapidly being enacted throughout the nation and will ultimately become part of the broad field of soil conservation.

SUMMARY

This chapter was presented to introduce the student to the broad areas of soil conservation, soil pollution and waste management, and land use regulation. Technology to handle soil conservation problems is readily available through the U.S. Department of Agriculture Soil Conservation Service and the Agricultural Extension Service. These same agencies can offer a great deal of assistance in solving soil pollution and waste management problems. It is your duty to become involved in the political process of land use regulation. It is our hope that you base your decisions on sound principles of soil science.

REVIEW QUESTIONS

1. Describe the four types of soil erosion.
2. Discuss the factors that control soil erosion.
3. Is wind erosion more important now than it was in 1900? Why or why not?
4. What are four methods of mechanical erosion control?
5. What vegetative erosion control measures are available?
6. When do agricultural chemicals pose a pollution threat to the environment?
7. How does animal waste management differ from industrial waste disposal?
8. How are soils important in urban and municipal waste management?
9. Are landfills a long-term solution to solid waste disposal? Why?
10. What are three methods of land use regulation? How does each method work?
11. Where can you obtain assistance in soil conservation and waste management problems?

12. The ultimate decisions involving land use belong to whom?

REFERENCES

Baver, L. D. *Soil Physics*. 3rd ed. New York: John Wiley, 1956, pp. 428-76.

Berger, Kermit C. *Introductory Soils*. New York: Macmillan Co., 1965, pp. 340-60.

Brady, Nyle C. *The Nature and Properties of Soils*. 8th ed. New York: Macmillan Co., 1974, pp. 551-92.

Donahue, Roy L., Shickluna, John C., and Robertson, Lynn S. *Soils: an Introduction to Soils and Plant Growth*. 3rd ed. Englewood Cliffs, N.J.: Prentice-Hall, 1971, pp. 366-78, 406-13, 471-86.

Foth, H. D., and Turk, L. M. *Fundamentals of Soil Science*. 5th ed. New York: John Wiley, 1972, pp. 345-74, 413-26.

McVickar, Malcolm H. *Using Commercial Fertilizers: Commercial Fertilizers and Crop Production*. 3rd ed. Danville, Ill.: Interstate Printers and Publishers, 1970, pp. 223-36.

Simonson, R.W., ed. *Non-Agricultural Applications of Soil Surveys, Developments in Soil Science 4*. Amsterdam: Elsevier Scientific Publishing Company, 1974.

Glossary

A horizon *See* soil horizon.

ABC soil A soil with a distinctly developed profile, including A, B, and C horizons.

absorption A term used in soils to mean that chemicals and fertilizers are taken into the soil.

AC soil A soil having a profile containing only A and C horizons with no clearly developed B horizon.

acid soil A soil with a preponderance of hydrogen ions, and probably of aluminum, in proportion to hydroxyl ions. Specifically, soil with a pH value of less than 7.0. For most practical purposes, a soil with a pH value of less than 6.6.

acidity, active The activity of hydrogen ion in the aqueous phase of a soil. It is measured and expressed as a pH value.

acidity, potential The amount of exchangeable hydrogen ion in a soil that can be rendered free or active in the soil solution by cation exchange. (Acidity on the soil colloids.)

This Glossary is reproduced from the Glossary of Soil Science Terms, Soil Science Society of America Proceedings, 29: 330-51, 1965, by permission of the Soil Science Society of America. *The 1957 Yearbook of Agriculture*, U.S. Department of Agriculture, Washington, D.C., was also used extensively in the compilation of this Glossary.

actinomycetes A general term applied to a group of organisms intermediate between the bacteria and the true fungi. They usually produce a characteristic branched mycelium and sporulate by segmentation of the entire mycelium or, more commonly, by segmentation of special terminal hyphae. Any organism belonging to the order of Actinomycetales.

adhesion In soils a term meaning a molecular attraction between soil colloids and water molecules or fertilizer ions.

adsorption The attachment of compounds or ionic parts of salts to a surface or another phase. Nutrients in solution (ions) carrying a positive charge become attached to (adsorbed by) negatively charged soil particles.

aerate To impregnate with a gas, usually air.

aeration, soil The process by which air in the soil is replaced by air from the atmosphere. In a well-aerated soil, the soil air is very similar in composition to the atmosphere above the soil. Poorly aerated soils usually contain a much higher percentage of carbon dioxide and a correspondingly lower percentage of oxygen than the atmosphere above the soil. The rate of aeration depends largely on the volume and continuity of pores within the soil.

aerobic 1. Having molecular oxygen as a part of environment. 2. Growing only in the presence of molecular oxygen, as aerobic organisms. 3. Occurring only in the presence of molecular oxygen (said of certain chemical or biochemical processes such as aerobic decomposition).

aggregate (of soil) Many fine soil particles held in a single mass or cluster, such as a clod, crumb, block, or prism. Many properties of the aggregate differ from those of an equal mass of unaggregated soil.

agronomy A specialization of agriculture concerned with the theory and practice of field-crop production and soil management. The scientific management of land.

air-dry 1. The state of dryness (of a soil) at equilibrium with the moisture content in the surrounding atmosphere. The actual moisture content will depend upon the relative humidity and the temperature of the surrounding atmosphere. 2. To allow to reach equilibrium in moisture content with the surrounding atmosphere.

air porosity The proportion of the bulk volume of soil that is filled with air at any given time or under a given condition, such as specified moisture tension. Usually the large pores· that is, those drained by a tension of less than approximately 100 cm of water.

alkali soil 1. A soil with a high degree of alkalinity (pH of 8.5 or higher) or with a high exchangeable sodium content (15 percent or more of the exchange capacity), or both. 2. A soil that contains sufficient alkali (sodium) to interfere with the growth of most crop plants.

alkaline soil Any soil whose pH is greater than 7.0.

alluvial fan An alluvial deposit at the base of a steep slope.

alluvial soil A soil developing from recently deposited alluvium and exhibiting essentially no horizon development or modification of the recently deposited materials.

alluvium Sand, mud, and other sediments deposited on land by streams.

amendment, soil 1. An alteration of the properties of a soil, and thereby of the soil, by the addition of substances such as lime, gypsum, sawdust, etc., for the purpose of making the soil more suitable for the production of plants. 2. Any such substance used for this purpose. Strictly speaking, fertilizers constitute a special group of soil amendments.

amino acids Amino acids are nitrogen-containing organic compounds, large numbers of which link together in the formation of a protein molecule. Each amino acid molecule contains one or more amino ($-NH_2$) groups and at least one carboxyl ($-COOH$) group. In addition, some amino acids (cystine and methionine) contain sulfur.

ammonification The biochemical process whereby ammoniacal nitrogen is released from nitrogen-containing organic compounds.

ammonium fixation The adsorption or absorption of ammonium ions by the mineral or organic fractions of the soil in a manner that they are relatively insoluble in water and relatively unexchangeable by the usual methods of cation exchange.

anaerobic 1. The absence of molecular oxygen. 2. Growing in the absence of molecular oxygen (such as anaerobic bacteria). 3. Occurring in the absence of molecular oxygen (as a biochemical process).

anion An ion carrying a negative charge of electricity.

anion-exchange capacity The sum total of exchangeable anions that a soil can adsorb. Expressed as milliequivalents per 100 grams of soil (or of other adsorbing material such as clay).

antibiotic A substance produced by one species of organism that, in low concentrations, will kill or inhibit growth of certain other organisms.

Ap horizon *See* soil horizon.

apatite A native phosphate of lime. The name is given to the chief mineral of phosphate rock and the inorganic compound of bones.

arid region Areas where the potential water losses by evaporation and transpiration are greater than the amount of water supplied by precipitation. In the United States this area is broadly considered to be the dry parts of the seventeen western states.

artificial manure *See* compost. (In European usage may denote commercial fertilizers.)

atom The amount of any element necessary for a chemical reaction.

autrophic Capable of utilizing carbon dioxide or carbonates as the sole source of carbon and obtaining energy for the reduction of carbon dioxide and for other life processes from the oxidation of inorganic elements or compounds such as iron, sulfur, hydrogen, ammonium, and nitrites, or from radiant energy.

available nutrient The portion of any element or compound in the soil that can be readily absorbed and assimilated by growing plants. ("Available" should not be confused with "exchangeable.")

available water The portion of water in a soil that can be readily absorbed by plant roots. Considered by most workers to be that water held in the soil against a pressure of up to approximately fifteen bars.

B horizon *See* soil horizon

bacteria Single-celled plants (microscopic).

banding of fertilizer The placement of fertilizers in the soil in continuous narrow ribbons, usually at specific distances from the seeds or plants. The fertilizer bands are covered by the soil but are not mixed with it.

bar A unit of pressure equal to one million dynes per square centimeter.

base-saturation percentage The extent to which the adsorption complex of a soil is saturated with exchangeable cations other than hydrogen. It is expressed as a percentage of the total cation-exchange capacity.

BC soil A soil profile with B and C horizons but with little or no A horizon.

bedding soil Arranging the surface of fields by plowing and grading into a series of elevated beds separated by shallow ditches for drainage.

bedrock The solid rock underlying soils and other earthy surface formations.

biological interchange The interchange of elements between organic and inorganic states in a soil or other substrate through the agency of biological activity. It results from biological decomposition of organic compounds and the liberation of inorganic materials (mineralization); and from the utilization of inorganic materials in the synthesis of microbial tissue (immobilization). Both processes commonly proceed continuously in soils.

biosequence A sequence of related soils that differ, one from the other, primarily because of differences in kinds and numbers of soil organisms as a soil-forming factor.

blowout An area from which soil material has been removed by wind. Such an area appears as a nearly barren, shallow depression with a flat or irregular floor consisting of a resistant layer, an accumulation of pebbles, or wet soil lying just above a water table.

bottomland *See* floodplain.

brackish Bodies of water which are only salty during dry periods, or water which is between fresh and sea water.

breccia A rock composed of coarse, angular fragments cemented together.

buffer compounds, soil The clay, organic matter, and compounds such as carbonates and phosphates which enable the soil to resist appreciable change in pH.

bulk density, soil The mass of dry soil per unit bulk volume. The bulk volume is determined before drying to the constant weight of 105C.

bulk specific gravity The ratio of the bulk density of a soil to the mass of unit volume of water.

bulk volume The volume, including the solids and the pores, of an arbitrary soil mass.

buried soil Soil covered by an alluvial, loessal, or other deposit, usually to a depth greater than the thickness of the solum.

C horizon *See* soil horizon.

calcareous soil Soil containing sufficient calcium carbonate (often with magnesium carbonate) to effervesce visibly when treated with cold 0.1N hydrochloric acid.

calcification (obsolete). The process or processes of soil formation in which the surface soil is kept sufficiently supplied with calcium to saturate the soil colloid, or the process of accumulation of calcium in some horizon of the profile.

calciphytes Plants that require or tolerate considerable amounts of calcium, or are associated with soils rich in calcium.

caliche 1. A layer near the surface, more or less cemented by secondary carbonates of calcium or magnesium precipitated from the soil solution. It may occur as a soft, thin soil horizon, as a hard, thick bed just beneath the solum, or as a surface layer exposed by erosion. Not a geologic deposit. 2. Alluvium cemented with sodium nitrate, chloride and/or other soluble salts in the nitrate deposits of Chile and Peru.

capillary fringe A zone just above the water table (zero gauge pressure) that remains almost saturated. (The extent and the degree of the capillary fringe depends upon the size-distribution of pores.)

capillary water (obsolete) The water held in the capillary or small pores of a soil, usually with a tension greater than 60 cm of water.

carbon cycle The sequence of transformation whereby carbon dioxide is fixed in living organisms by photosynthesis or by chemosynthesis, liberated by respiration and by the death and decomposition of the fixing organism, used by heterotrophic species, and ultimately returned to its original state.

carbon-nitrogen ratio The ratio of the weight of organic carbon to the weight of total nitrogen in a soil or in organic material. It is obtained by dividing the percentage of organic carbon (C) by the percentage of total nitrogen (N).

category Any one of the ranks of the system of soil classification in which soils are grouped on the basis of their characteristics.

catena A sequence of soils of about the same age, derived from similar parent material, and occurring under similar climatic conditions, but having different characteristics due to variation in relief and in drainage.

cation An ion carrying a positive charge of electricity. The common soil cations are calcium, magnesium, sodium, potassium, aluminum, and hydrogen.

cation exchange The interchange between a cation in solution and another cation on the surface of any surface-active material such as clay colloid or organic colloid.

cation-exchange capacity The sum total of exchangeable cations that a soil can absorb. Sometimes called total-exchange capacity, base-exchange capacity, or cation-adsorption capacity. Expressed in milliequivalents per 100 grams of soil (or of other adsorbing material such as clay).

cemented Indurated; having a hard, brittle consistency because the particles are held together by cementing substances such as humus, calcium carbonate, or the oxides of silicon, iron, and aluminum. The hardness and brittleness persist even when wet.

chemically precipitated phosphorus Relatively insoluble phosphorus compounds resulting from reactions between constituents in solution to form chemically homogeneous particles of the solid phase. Examples are: calcium and magnesium phosphates which are precipitated above a pH of about 6.0 to 6.5 (if calcium and magnesium are present); and, iron and aluminum phosphates which are precipitated below a pH of about 5.8 to 6.1 at which many iron and aluminum compounds are soluble. A form of fixed phosphate.

chemisorbed phosphorus Phosphorus adsorbed or precipitated on the surface of clay minerals or other crystalline materials as a result of the attractive forces between the phosphate ion and constituents in the surface of the solid phase.

chisel To break or shatter compact soil or subsoil layers by use of a chisel.

chisel, subsoil A tillage implement with one or more cultivator-type feet to which are attached strong knifelike units used to shatter or loosen hard, compact layers, usually in the subsoil, to depths below normal plow depth.

chlorosis A condition in plants resulting from the failure of chlorophyll (the green coloring matter) to develop, usually because of deficiency of an essential nutrient. Leaves of chlorotic plants range from light green through yellow to almost white.

chroma The relative purity, strength, or saturation of a color; directly related to the dominance of the determining wavelength of the light and inversely related to grayness; one of the three variables of color.

chronosequence A sequence of related soils that differ, one from the other, in certain properties primarily as a result of time as a soil-forming factor.

class, soil A group of soils having a definite range in a particular property such as acidity, degree of slope, texture, structure, land use capability, degree of erosion, or drainage. *See* soil texture and soil structure.

classification, soil The systematic arrangement of soils into groups or categories on the basis of their characteristics.

clay 1. A soil separate consisting of particles less than or equal to 0.002 mm in equivalent diameter. 2. A textural class.

clayey Containing large amounts of clay or having properties similar to those of clay.

clay loam A textural class. *See* soil texture.

clay mineral 1. Naturally occurring, inorganic crystalline material found in soils and other earthy deposits, the particles being of clay size; that is, less than or equal to 0.002 mm in diameter. 2. Material as described under 1, but not limited by particle size.

clay pan A dense, compact layer in the subsoil having a much higher clay content than the overlying material, from which it is separated by a sharply defined boundary; formed by downward movement of clay or by synthesis of clay in place during soil formation. Clay pans are usually hard when dry, and plastic and sticky when wet. Also, they usually impede the movement of water and air, and the growth of plant roots.

clod A mass of soil produced by plowing or digging, which usually slakes easily with repeated wetting and drying, in contrast to a ped, which is a natural soil aggregate.

coarse fragments Rock or mineral particles greater than 2.0 mm in diameter.

coarse texture The texture exhibited by sands, loamy sands, and sandy loams except very fine, sandy loam. A soil containing large quantities of these textural classes.

cobblestone Rounded or partially rounded rock or mineral fragments between 3 and 10 inches in diameter.

cohesion The force holding two like substances together.

colloid, soil Colloid refers to organic or inorganic matter having very small-particle size and a correspondingly large surface area per unit of mass. Most colloidal particles are too small to be seen with the ordinary compound microscope. Soil colloids do not go into true solution as sugar or salt do, but they may be dispersed into a relatively stable suspension and thus be carried in moving water. By treatment with salts and other chemicals, colloids may be flocculated, or aggregated, into small crumbs or granules that settle out of water. (Such small crumbs of aggregated colloids can be moved by rapidly moving water or air just as other particles can be.) Many mineral soil colloids are really tiny crystals and the minerals can be identified with X-rays and in other ways.

colluvium A deposit of rock fragments and soil material accumulated at the base of steep slopes as a result of gravitational action.

compaction, soil Compressing soils such that air space is minimized.

compost A mass of rotted organic matter made from waste plant residues. Inorganic fertilizers, especially nitrogen, and a little soil usually are added to it. The organic residues usually are piled in layers, to which the fertilizers are added. The layers are separated by thin layers of soil. The whole pile is kept moist and allowed to decompose. The pile is usually turned once or twice. The principal purpose in making compost is to permit the organic materials to become crumbly and to reduce the carbon-nitrogen ratio of the material. Compost is sometimes called artificial or synthetic manure.

concretion A local concentration of a chemical compound, such as calcium carbonate or iron oxide, in the form of a grain or nodule of varying size, shape, hardness, and color.

consistence The combination of properties of soil material that determine its resistance to crushing and its ability to be molded or changed in shape. Consistence depends mainly on the forces of attraction between soil particles. Consistence is described by such words as loose, friable, firm, soft, plastic, and sticky.

consistency 1. The resistance of a material to deformation or rupture. 2. The degree of cohesion or adhesion of the soil mass. Terms used for describing consistency at various soil moisture contents are:

wet soil—nonsticky, slightly sticky, sticky, very sticky, nonplastic, slightly plastic, plastic, and very plastic.

moist soil—loose, very friable, friable, firm, very firm, and extremely firm.

dry soil—loose, soft, slightly hard, hard, very hard, and extremely hard.

cementation—weakly cemented, strongly cemented, and indurated.

consolidate To place into a compact mass and thus increase density and reduce pore space.

contour An imaginary line connecting points of equal elevation on the surface of the soil. A contour terrace is laid out on a sloping soil at right angles to the direction of the slope and level throughout its course. In contour plowing, the plowman keeps to a level line at right angles to the direction of the slope, which usually results in a curving furrow.

cover crop A crop planted on cultivated soils to prevent erosion during periods of nonproduction.

creep The downward mass movement of sloping soil. The movement is usually slow and irregular and occurs most commonly when the lower soil is nearly saturated with water.

crop residue The unharvested portions of crops.

crop rotation The cycling sequence of several crops grown over a period of years on a given piece of land

crotovina A former animal burrow in one soil horizon that has been filled with organic matter or material from another horizon.

crumb structure A structural condition in which most of the peds are crumbs.

crust A thin, brittle layer of hard soil that forms on the surface of many soils when they are dry. An exposed, hard layer of materials cemented by calcium carbonate, gypsum, or other binding agents. Most desert crusts are formed by the exposure of such layers through removal of the upper soil by wind or running water and their subsequent hardening.

crystal A homogeneous, inorganic substance of definite chemical composition bounded by plane surfaces that form definite angles with each other, thus giving the substance a regular geometrical form. *See* soil mineral.

crystalline rock A rock consisting of various minerals that have crystallized in place from magma. *See* igneous rock and sedimentary rock.

cultivation A tillage operation used in preparing land for seeding or transplanting, or later for weed control and for loosening the soil.

damping-off Sudden wilting and death of seedling plants resulting from attack by microorganisms.

decalcification The removal of calcium carbonate or calcium ions from the soil by leaching.

decomposition *See* mineralization.

deflocculate To separate or to break up soil aggregates into the individual particles; to disperse the particles of a granulated clay to form a clay that runs together or puddles.

degradation (obsolete) The changing of a soil to a more highly leached or more highly weathered condition, usually accompanied by morphological changes such as development of an A2 horizon.

delta An alluvial deposit formed where a river drops alluvial materials when it flows into the sea.

denitrification The biochemical reduction of nitrate or nitrite to gaseous nitrogen either as molecular nitrogen or as an oxide of nitrogen.

deposit Material left in a new position by a natural transporting agent such as water, wind, ice, or gravity, or by the activity of man.

desalination Removal of salts from saline soil, usually by leaching.

desert crust A hard layer, containing calcium carbonate, gypsum, or other binding material, exposed at the surface in desert regions.

desert pavement The layer of gravel or stones left on the land surface in desert regions after the removal of the fine material by wind erosion.

desorption The removal of adsorbed materials from surfaces.

diatoms Algae having siliceous cell walls that persist as a skeleton after death. Any of the microscopic unicellular or colonial algae constituting the class Bacillariaceae. They occur abundantly in fresh and salt waters and their remains are widely distributed in soils.

diatomaceous earth A geologic deposit of fine, grayish siliceous material composed chiefly or wholly of the remains of diatoms. It may occur as a powder or as a porous, rigid material.

diffusion The transport of matter as a consequence of the movement of the constituent particles. The intermingling of two gases or liquids in contact with each other takes place by diffusion.

disintegration *See* physical weathering.

disperse 1. To break up compound particles, such as aggregates, into the individual component particles. 2. To distribute or suspend fine particles, such as clay, in or throughout a dispersion medium, such as water.

diversion dam A structure or barrier built to divert part or all of the water of a stream to a different course.

double layer In colloid chemistry, the electric charges on the surface of the disperse phase (usually negative), and the adjacent diffuse layer (usually positive) of ions in solution.

drain 1. To provide channels, such as open ditches or drain tile, so that excess water can be removed by surface or by internal flow. 2. To lose water (from the soil) by percolation.

drainage The removal of excess surface water or excess water from within the soil by means of surface or subsurface drains.

drainage, excessive Too great or too rapid loss of water from soils, either by percolation or by surface flow. Loss greater than that necessary to prevent the development of an anaerobic condition for any appreciable length of time.

drainage, soil 1. The rapidity and extent of the removal of water from the soil by runoff and flow through the soil to underground spaces. 2. As a condition of the soil, soil drainage refers to the frequency and duration of periods when the soil is free of saturation. For example, in well-drained soils, the water is removed readily, but not rapidly; in poorly drained soils, the root zone is waterlogged for long periods and the roots of ordinary crop plants cannot get enough oxygen; and in excessively drained soils, the water is removed so completely that most crop plants suffer from lack of water.

drain tile Concrete or ceramic pipe used to conduct water from the soil.

drift Material of any sort deposited by geological processes in one place after having been removed from another. Glacial drift includes any materials deposited by glaciers, and by the streams and lakes associated with them.

drought (drouth) A period of dryness, especially a long one. Usually considered to be any period of soil moisture deficiency within the plant root zone. A period of dryness of sufficient length to deplete soil moisture to the extent that plant growth is seriously retarded.

dry-weight percentage The ratio of the weight of any constituent (of a soil) to the oven-dry weight of the soil. *See* oven-dry soil.

dust mulch A loose, finely granular, or powdery condition on the surface of the soil, usually produced by shallow cultivation.

dynamometer An instrument for measuring draft of tillage implements and for measuring resistance of soil to penetration by tillage implements.

ecology The branch of biology that deals with the mutual relations among organisms and between organisms and their environment.

ectotrophic mycorrhiza A mycorrhizal association in which the fungal hyphae form a compact mantle on the surface of the roots. Mycelial strands extend inward between cortical cells and outward from the mantle to the surrounding soil.

edaphic 1. Of or pertaining to the soil. 2. Resulting from or influenced by factors inherent in the soil or other substrate, rather than by climatic factors.

edaphology The science that deals with the influence of soils on living things, particularly plants, including man's use of land for plant growth.

effective precipitation That portion of the total precipitation which becomes available for plant growth.

effluent The outflowing of water from a subterranean storage space. (Also used generally for gases and other liquids.)

eluvial horizon A soil horizon that has been formed by the process of eluviation.

eluviation The movement of material from one place to another within the soil in either true solution or colloidal suspension. Soil horizons that have lost material through eluviation are said to be eluvial; those that have received material are illuvial. With an excess of rainfall over evaporation, eluviation may take place either downward or laterally according to the direction of water movement. The term refers especially to the movement of soil colloids in suspension; leaching refers to the removal of soluble materials such as salt in true solution.

endodynamomorphic soils Soils with properties that have been influenced primarily by parent material.

endotrophic Nourished or receiving nourishment from within, as fungi or their hyphae receiving nourishment from plant roots in a mycorrhizal association.

endotrophic mycorrhiza A mycorrhizal association in which the fungal hyphae are present on root surfaces only as individual threads that may penetrate directly into root hairs, other epidermal cells and occasionally into cortical cells. Individual threads extend from the root surface outward into the surrounding soil.

environment All external conditions that may act upon an organism or soil to influence its development, including sunlight, temperature, moisture, and other organisms.

enzymes Substances produced by living cells which can bring about or speed up chemical reaction. They are organic catalysts.

epipedon A diagnostic surface horizon in the modern soil classification system.

equivalent weight of a soil colloid The weight of a clay or organic colloid that has a combining power equivalent to 1 gram atomic weight of hydrogen.

erode To wear away or remove the land surface by wind, water, or other agents.

erodible Susceptible to erosion.

erosion 1. The wearing away of the land surface by running water, wind, ice or other geological agents. 2. Detachment and movement of soil or rock by water, wind, ice, or gravity.

erosion classes A grouping of erosion conditions based on the degree of erosion or on characteristic patterns.

eutrophic Having concentrations of nutrients optimal (or nearly so) for plant or animal growth. (Said of nutrient solutions or of soil solutions.)

evapotranspiration The loss of water from a soil by evaporation and plant transpiration.

exchange acidity The titratable hydrogen and aluminum that can be replaced from the adsorption complex by a neutral salt solution. Usually expressed as milliequivalents per 100 grams of soil.

exchange capacity The total ionic charge of the adsorption complex active in the adsorption of ions.

exchangeable-cation percentage The extent to which the adsorption complex of a soil is occupied by a particular cation. It is expressed as follows:

$$\mathrm{ECP} = \frac{\text{Exchangeable cation (meq/100 g soil)}}{\text{Cation-exchange capacity (meq/100 g soil)}} \times 100.$$

exchangeable potassium The potassium that is held by the adsorption complex of the soil and is easily exchanged with the cation of neutral nonpotassium salt solutions.

exchangeable-sodium percentage The extent to which the adsorption complex of a soil is occupied by sodium. It is expressed as follows:

$$ESP = \frac{\text{Exchangeable sodium (meq/100 g soil)}}{\text{Cation-exchange capacity (meq/100 g soil)}} \times 100.$$

fallow Cropland left idle in order to restore productivity, mainly through accumulation of water, nutrients, or both. Summer fallow is a common stage before cereal grain in regions of limited rainfall. The soil is tilled for at least one growing season to control weeds, to aid decomposition of plant residues, and to encourage the storage of moisture for the succeeding grain crop. Bush or forest fallow is a rest period under woody vegetation between crops.

family, soil In soil classification, it is a category between the subgroup and the soil series.

fertility, soil The status of a soil with respect to the amount and availability to plants of elements necessary for plant growth.

fertilizer Any organic or inorganic material of natural or synthetic origin which is added to a soil to supply certain elements essential to the growth of plants.

fertilizer grade The guaranteed minimum analysis, in percent, of the major plant nutrient elements contained in a fertilizer material or in a mixed fertilizer. (Usually refers to the percentage of $N\text{-}P_2O_5\text{-}K_2O$, but proposals are pending to change the designation to the percentage of N-P-K.)

field capacity (field moisture capacity) (obsolete in technical work) The percentage of water remaining in a soil two or three days after having been saturated and after free drainage has practically ceased. (The percentage may be expressed on the basis of weight or volume.)

fine texture Soils consisting of or containing large quantities of the fine fractions, particularly of silt and clay.

first bottom The normal floodplain of a stream.

fixation The process or processes in a soil by which certain chemical elements essential for plant growth are converted from a soluble or exchangeable form to a much less soluble or to a nonexchangeable form.

fixed phosphorus That phosphorus which has been changed to a less soluble form as a result of reaction with the soil.

flocculate To aggregate or clump together individual tiny soil particles, especially fine clay, into small groups or granules. The opposite of deflocculate, or disperse.

floodplain The land bordering a stream, built up of sediments from overflow of the stream, and subject to inundation when the stream is at flood state. *See* first bottom.

foliar fertilization Fertilization of plants by applying chemical fertilizers to their foliage.

forest floor All dead vegetable or organic matter, including litter and unincorporated humus, on the mineral soil surface under forest vegetation.

forest soils Soils developed under forest vegetation.

fragipan A natural subsurface horizon with high bulk density relative to the solum above, seemingly cemented when dry, but when moist showing a moderate to weak brittleness.

friable A consistency term pertaining to the ease of crumbling of soils.

fungi Forms of plant life lacking chlorophyll and unable to make their own food.

genesis, soil The mode of origin of the soil, with special reference to the processes responsible for the development of the solum, or true soil, from the unconsolidated parent material.

glacial drift Rock debris that has been transported by glaciers and deposited, either directly from the ice or from the melt-water. The debris may or may not be heterogeneous.

glacial till *See* till.

Gley soil (obsolete in the U.S.) Soil developed under conditions of poor drainage resulting in reduction of iron and other elements and in gray colors and mottles.

granular structure Soil structure in which the individual grains are grouped into spherical aggregates with indistinct sides. Highly porous granules are commonly called crumbs. A well-granulated soil has the best structure for most ordinary crop plants.

gravitational water Water which moves into, through, or out of the soil under the influence of gravity.

great group A category in soil classification between the suborder and subgroup; concerned with presence of major soil horizons.

green manure Plant material incorporated with the soil while green, or soon after maturity, for improving the soil.

ground water That portion of the total precipitation which at any particular time is either passing through or standing in the soil and the underlying strata and is free to move under the influence of gravity.

gully A channel resulting from erosion and caused by the concentrated but intermittent flow of water usually during, and immediately following, heavy rains. Deep enough to interfere with, and not to be obliterated by, normal tillage operations.

halomorphic soil Soil formed under imperfect drainage in arid regions.

hardpan A hardened soil layer, in the lower A or in the B horizon, caused by cementation of soil particles with organic matter, or with materials like silica, sesquioxides, or calcium carbonate. The hardness does not change appreciably with changes in moisture content, and pieces of the hard layer do not slake in water.

Histosols A group of soils developed from organic materials.

horizon *See* soil horizon.

hue One of the three variables of soil color. It is caused by light of certain wavelengths and changes with the wavelength.

humification The processes involved in the decomposition of organic matter and leading to the formation of humus.

humus That more or less stable fraction of the soil organic matter remaining after the major portion of added plant and animal residues have decomposed. Usually it is dark in color.

hydrologic cycle The fate of water from the time of precipitation until the water has been returned to the atmosphere by evaporation and is again ready to be precipitated.

hydromorphic soils Soils formed under conditions of poor drainage in marshes, swamps, seepage areas, or flats.

hygroscopic Capable of taking up moisture from the air.

hygroscopic water Water adsorbed by a dry soil from an atmosphere of high relative humidity, water remaining in the soil after "air-drying," or water held by the soil when it is in equilibrium with an atmosphere of a specified relative humidity at a specified temperature, usually 98 percent relative humidity at 25C.

igneous rock Rock formed from the cooling and solidification of magma, and that has not been changed appreciably since its formation.

illite A hydrous mica.

illuviation The process of deposition of soil material removed from one horizon to another in the soil; usually from an upper to a lower horizon in the soil profile. *See* eluviation.

immobilization The conversion of an element from the inorganic to the organic form in microbial tissues or in plant tissues, thus rendering the element not readily available to other organisms or to plants.

impeded drainage A condition which hinders the movement of water through soils under the influence of gravity.

impervious Resistant to penetration by fluids or by roots.

Inceptisols Young soils showing only beginning soil formation.

infiltration The downward entry of water into the soil.

infiltration rate A soil characteristic determining or describing the maximum rate at which water can enter the soil under specified conditions.

inorganic Refers to substances occurring as minerals in nature or obtainable from them by chemical means. Refers to all matter except the compounds of carbon, but includes carbonates.

inorganic nitrogen Nitrogen in combination with mineral elements, not in animal or vegetable form. Ammonium sulfate and sodium nitrate are examples of inorganic nitrogen combinations, while proteins contain nitrogen in organic combination.

intergrade A soil that possesses moderately well-developed distinguishing characteristics of two or more genetically related soil groups.

ions Atoms, groups of atoms, or compounds, which are electrically charged as a result of the loss of electrons (cations) or the gain of electrons (anions).

iron pan An indurated soil horizon in which iron oxide is the principal cementing agent.

irrigation The artificial application of water to the soil for the benefit of growing crops.

irrigation methods The manner in which water is artifically applied to an area. The methods and the manner of applying the water are as follows:

border-strip—The water is applied at the upper end of a strip with earth borders to confine the water to the strip.

flooding—The water is released from field ditches and allowed to flood over the land.

furrow—The water is applied to row crops in ditches made by tillage implements.

sprinkler—The water is sprayed over the soil surface through nozzles from a pressure system.

isomorphous substitution The replacement of one atom by another of similar size in a crystal lattice without disrupting or changing the crystal structure of the mineral.

isotope One of two or more forms of a chemical element having the same atomic number and position in the periodic table of elements, but distinguishable by differences of weight.

kame An irregular ridge or hill of stratified glacial drift.

kaolin 1. An aluminosilicate mineral of the 1:1 crystal lattice group; that is, consisting of one silicon tetrahedral layer and one aluminum oxide-hydroxide octahedral layer. 2. the 1:1 group or family of aluminosilicates.

lacustrine deposit Material deposited in lake water and later exposed either by lowering of the water level or by the elevation of the land.

land The total natural and cultural environment within which production takes place. Land is a broader term than soil. In addition to soil, its attributes include other physical conditions such as mineral deposits and water supply; location in relation to centers of commerce, populations, and other land; the size of the individual tracts or holdings; and existing plant cover, works of improvement, and the like. Some use the term loosely in other senses: As defined above, but without the economic or cultural criteria, especially in the expression *natural land*; as a synonym for *soil*; for the solid surface of the earth; and also for earthy surface formations, especially in the geomorphological expression *land form*.

land-capability classification A grouping of kinds of soil into special units, subclasses, and classes according to their capability for intensive use and the treatments required for sustained use.

land classification The arrangement of land units into various categories based upon the properties of the land or its suitability for some particular purpose.

landscape All the natural features such as fields, hills, forests, water, etc., which distinguish one part of the earth's surface from another part. Usually that portion of land or territory which the eye can comprehend in a single view, including all its natural characteristics.

land use planning The development of plans for the uses of land that, over long periods, will best serve the general welfare, together with the formulation of ways and means for achieving such uses.

leaching The removal of materials in solution from the soil.

lignin An organic substance that incrusts the cellulose framework of plant cell walls. It is made up of modified phenyl propane units. It is dissolved only with difficulty and is more inert chemically and biologically than other plant constituents. Lignin increases with age in plants.

liquid limit The minimum percentage (by weight) of moisture at which a small sample of soil will barely flow under a standard treatment.

lithosequence A group of related soils that differ, one from the other, in certain properties primarily as a result of differences in the parent rock as a soil-forming factor.

loam A soil textural class.

loamy Intermediate in texture and properties between fine-textured and coarse-textured soils. Includes all textural classes with the words *loam* or *loamy* as a part of the class name, such as clay loam or loamy sand.

loess Material transported and deposited by wind and consisting of predominantly silt-sized particles.

luxury consumption The intake by a plant of an essential nutrient in amounts exceeding what it needs. Thus if potassium is abundant in the soil, alfalfa may take in more than is required.

macronutrient A chemical element necessary in large amounts (usually greater than 1 ppm in the plant) for the growth of plants and usually applied artificially in fertilizer or liming materials (macro refers to quantity and not the essentiality of the element).

manure The excreta of animals, with or without the admixture of bedding or litter, in varying stages of decomposition. *Also referred to as* barnyard manure or stable manure.

marl Soft and unconsolidated calcium carbonate, usually mixed with varying amounts of clay or other impurities.

marsh Periodically wet or continually flooded areas with the surface not deeply submerged. Covered dominantly with sedges, cattails, rushes, or other hydrophytic plants. Subclasses include fresh water and salt water marshes.

mature soil A soil with well-developed soil horizons produced by the natural processes of soil formation and essentially in equilibrium with its present environment.

maximum water-holding capacity The average moisture content of a disturbed sample of soil, 1 cm high, which is at equilibrium with a water table at its lower surface.

mechanical analysis (obsolete) *See* particle-size analysis and particle-size distribution.

medium-texture Intermediate between fine-textured and coarse-textured (soils). (It includes the following textural classes: very fine sandy loam, loam, silt loam, and silt.)

mellow soil A very soft, friable, porous soil without any tendency toward hardness or harshness.

metamorphic rock Rock derived from preexisting rocks but differing from them in physical, chemical, and mineralogical properties as a result of natural geological processes, principally heat and pressure, originating within the earth. The preexisting rocks may have been igneous, sedimentary, or another form of metamorphic rock.

micas Primary alumino-silicate minerals in which two silica layers alternate with one alumina layer. They separate readily into thin sheets or flakes.

microclimate 1. The climatic condition of a small area resulting from the modification of the general climatic conditions by local differences in elevation or exposure. 2. The sequence of atmospheric changes within a very small region.

microfauna That part of the animal population which consists of individuals too small to be clearly distinguished without the use of a microscope. Includes protozoa and nematodes.

microflora The part of the plant population which consists of individuals too small to be clearly distinguished without the use of a microscope. Includes actinomycetes, algae, bacteria, and fungi.

micronutrient A chemical element necessary in only extremely small amounts (less than 1 ppm in the plant) for the growth of plants. Examples are B, Cl, Cu, Fe, Mn, and Zn. (Micro refers to the amount used rather than to its essentiality.)

microrelief Small-scale or local differences in topography, including mounds, swales, or pits that are only a few feet in diameter and with elevation differences of up to 6 feet.

mineralization The conversion of an element from an organic form to an inorganic state as a result of microbial decomposition.

mineral soil A soil consisting predominantly of, and having its properties determined predominantly by, mineral matter. Usually contains less than 20 percent organic matter, but may contain an organic surface layer up to 30 cm thick.

minor element (obsolete) *See* micronutrient.

moisture equivalent The weight percentage of water retained by a previously saturated sample of soil 1 cm in thickness after it has been subjected to a centrifugal force of 1,000 times gravity for 30 minutes.

moisture-retention curve A graph showing the soil moisture percentage (by weight or by volume) versus applied tension (or pressure). Points on the graph are usually obtained by increasing (or decreasing) the applied tension or pressure over a specified range.

moisture tension (or pressure) The equivalent negative pressure in the soil water. It is equal to the equivalent pressure that must be applied to the soil water to bring it to hydraulic equilibrium, through a porous permeable wall or membrane, with a pool of water of the same composition.

montmorillonite An aluminosilicate clay mineral with a 2:1 expanding crystal lattice; that is with two silicon tetrahedral layers enclosing an aluminum octahedral layer. Considerable expansion may be caused along the C axis by water moving between silica layers of continuous units. *See* montmorillonite group.

montmorillonite group Clay minerals with 2:1 crystal lattice structure, that is, two silicon tetrahedral layers enclosing an aluminum octahedral layer. Consists of montmorillonite, beidellite, nontronite, saponite, and others. This group is commonly called smectites.

moraine Accumulation of soil materials left by glaciers.

morphology, soil The composition of the soil including texture, structure, consistence, color, and other physical, chemical, and biological properties of the various soil horizons that make up the soil profile.

mottling Spots or blotches of different color or shades of color interspersed with the dominant color.

muck Highly decomposed organic material in which the original plant parts are not recognizable. Contains more mineral matter and is usually darker in color than peat.

mulch 1. Any material such as straw, sawdust, leaves, plastic film, loose soil, etc., that is spread upon the surface of the soil to protect the soil and plant roots from the effects of raindrops, soil crusting, freezing, evaporation, etc. 2. To apply mulch to the soil surface.

Munsell color system A color designation system that specifies the relative degrees of the three simple variables of color: hue, value, and chroma. For example: 10YR 6/4 is a color (of soil) with a hue = 10YR, value = 6, and chroma = 4. These notations can be translated into several different systems of color names as desired.

mycorrhiza The association, usually symbiotic, of fungi with the roots of seed plants.

nematodes Very small worms abundant in many soils and important because many of them attack and destroy plant roots.

neutral soil A soil in which the surface layer, at least to normal plow depth, is neither acid nor alkaline in reaction.

nitrate reduction The biochemical reduction of nitrate.

nitrification The biochemical oxidation of ammonium to nitrate.

nitrogen assimilation The incorporation of nitrogen into organic cell substances by living organisms.

nitrogen cycle The sequence of biochemical changes undergone by nitrogen wherein it is used by a living organism, liberated upon the death and decomposition of the organism, and converted to its original state of oxidation.

nitrogen fixation The conversion of elemental nitrogen (N_2) to organic combination or to forms readily utilizable in biological processes.

nodule bacteria *See* rhizobia.

nutrient, plant Any element taken in by a plant, essential to its growth, and used by it in elaboration of its food and tissue.

O horizon *See* soil horizon.

order The highest category in soil classification.

organic phosphorus Phosphorus present as a constituent of an organic compound or a group of organic compounds, such as glycerophosphoric acid, inositol phosphoric acid, cytidlyic acid, etc.

organic soil A soil which contains a high percentage (greater than 15 or 20 percent) of organic matter throughout the solum.

osmotic A type of pressure exerted in living bodies as a result of unequal concentration of salts on both sides of a cell wall or membrane. Water will move from the area having the least salt concentration through the membrane into the area having the highest salt concentration and, therefore, exerts additional pressure on this side of the membrane.

oven-dry soil Soil which has been dried at 105°C until it reaches a constant weight.

Oxisols The high iron and aluminum soils of the tropics.

pans Horizons or layers, in soils, that are strongly compacted, indurated, or very high in clay content.

parent material The unconsolidated and more or less chemically weathered mineral or organic matter from which the solum of soils is developed by pedogenic processes.

particle density The mass per unit volume of the soil particles. In technical work, usually expressed as grams per cubic centimeter.

particle size The effective diameter of a particle measured by sedimentation, sieving, or micrometric methods.

particle-size analysis Determination of the various amounts of the different separates in a soil sample, usually by sedimentation, micrometry, sieving, or combinations of these methods.

particle-size distribution The amounts of the various soil separates in a soil sample, usually expressed as weight percentages.

parts per million (ppm) Weight units of any given substance per one million equivalent weight units of oven-dry soil; or, in the case of soil solution or other solution, the weight units of solute per million weights of solution.

peat Unconsolidated soil material consisting largely of undecomposed, or only slightly decomposed, organic matter accumulated under conditions of excessive moisture.

ped A unit of soil structure such as an aggregate, crumb, prism, block, or granule, formed by natural processes (in contrast with a clod, which is formed artificially.)

pedalfer (obsolete) A subdivision of a soil order comprising a large group of soils in which sesquioxides increased relative to silica during soil formation.

pedocal (obsolete) A subdivision of a soil order comprising a large group of soils in which calcium accumulated during soil formation.

pedon A three-dimension soil body depicting the range of characteristics for a given soil.

percolation, soil water The downward movement of water through soil. Especially, the downward flow of water in saturated or nearly saturated soil at hydraulic gradients of the order of 1.0 or less.

permafrost 1. Permanently frozen material underlying the solum. 2. A perennially frozen soil horizon.

permanent charge The net negative (or positive) charge of clay particles inherent in the crystal lattice of the particle; not affected by changes in pH or by ion-exchange reactions.

permeability, soil 1. The ease with which gases, liquids, or plant roots penetrate or pass through a bulk mass of soil or a layer of soil. Since different soil horizons vary in permeability, the particular horizon under question should be designated. 2. The property of a porous medium itself that relates to the ease with which gases, liquids, or other substances can pass through it.

pH, soil The negative logarithm of the hydrogen-ion activity of a soil. The degree of acidity (or alkalinity) of a soil as determined by means of a glass, quinhydrone, or other suitable electrode or indicator at a specified moisture content or soil-water ratio, and expressed in terms of the pH scale.

pH-dependent charge That portion of the total charge of the soil (clay) particles which is affected by, and varies with, changes in pH.

phase, soil A subdivision of a soil series.

photosynthesis The process of conversion by plants of water and carbon dioxide into carbohydrates under the action of light. Chlorophyll is required for the conversion of the light energy into chemical forms.

physical properties (of soils) Those characteristics, processes, or reactions of a soil which are caused by physical forces and which can be described by, or expressed in, physical terms or equations. Sometimes confused with and difficult to separate from chemical properties; hence, the terms physical-chemical or physiochemical. Examples of physical properties are bulk density, water-holding capacity, hydraulic conductivity, porosity, pore-size distribution, etc.

physical weathering The breakdown of rock and mineral particles into smaller particles by physical forces such as frost action. *See* weathering.

plastic limit The minimum moisture percentage by weight at which a small sample of soil material can be deformed without rupture. *Synonymous with* lower plastic limit.

polypedon A soil classification term referring to a group of pedons with similar characteristics.

pore space Total space not occupied by soil particles in a bulk volume of soil.

porosity The volume percentage of the total bulk not occupied by solid particles.

potassium fixation The process of converting exchangeable or water-soluble potassium to moderately soluble potassium, i.e., to a form not easily exchanged from the adsorption complex with a cation of a neutral salt solution.

primary mineral A mineral that has not been altered chemically since deposition and crystallization from molten lava.

prismatic soil structure A soil structure type with prismlike aggregates that have a vertical axis much longer than the horizontal axes.

productivity, soil The capacity of a soil, in its normal environment, for producing a specified plant or sequence of plants under a specified system of management.

profile, soil A vertical section of the soil through all its horizons and extending into the parent material.

protein Any of a group of nitrogen-containing compounds that yield amino acids on hydrolysis and have high molecular weights. They are essential parts of living matter and are one of the essential food substances of animals.

puddled soil Dense, massive soil artificially compacted when wet and having no regular structure. The condition commonly results from the tillage of a clayey soil when it is wet.

reaction, soil The degree of acidity or alkalinity of a soil, usually expressed as a pH value.

regolith The unconsolidated mantle of weathered rock and soil material on the earth's surface; loose earth materials above solid rock. (Approximately equivalent to the term *soil* as used by many engineers.)

relief Elevations or inequalities of the land surface, considered collectively.

residual material Unconsolidated and partly weathered mineral materials accumulated by disintegration of consolidated rock in place.

rhizobia Bacteria capable of living symbiotically with higher plants, usually legumes, from which they receive their energy, and capable of using atmospheric nitrogen; hence, the term symbiotic nitrogen-fixing bacteria. (Derived from the generic name *Rhizobium*.)

rill A small, intermittent water course with steep sides; usually only a few inches deep and, hence, no obstacle to tillage operations.

rock A complex mineral aggregate.

root zone The part of the soil that is invaded by plant roots.

runoff That portion of the precipitation on an area which is discharged from the area through stream channels. That which is lost without entering the soil is called surface runoff and that which enters the soil before reaching the stream is ground-water runoff or seepage flow from ground water. (In soil science runoff usually refers to the water lost by surface flow; in geology and hydraulics runoff usually includes both surface and subsurface flow.)

sand 1. A soil particle between 0.05 and 2.0 mm in diameter. 2. Any one of five soil separates, namely: very coarse sand, coarse sand, medium sand, fine sand, and very fine sand. 3. A soil textural class.

sandy Containing a large amount of sand. (Applied to any one of the soil classes that contains a large percentage of sand.)

sandy clay A soil textural class.

sandy clay loam A soil textural class.

sandy loam A soil textural class.

saturate 1. To fill all the voids between soil particles with a liquid. 2. To form the most concentrated solution possible under a given set of physical conditions in the presence of an excess of the solute. 3. To fill to capacity, as the adsorption complex with a cation species, e.g., H-saturated, etc.

second bottom The first terrace above the normal floodplain of a stream.

secondary mineral A mineral resulting from the decomposition of a primary mineral or from the reprecipitation of the products of decomposition of a primary mineral.

sedimentary rock A rock formed from materials deposited from suspension or precipitated from solution and usually being more or less consolidated. The principal sedimentary rocks are sandstones, shales, limestones, and conglomerates.

self-mulching soil A soil in which the surface layer becomes so well aggregated that it does not crust and seal under the impact of rain but instead serves as a surface mulch upon drying.

shear Force, as of a tillage implement, acting at right angles to the direction of movement.

silica An important soil constituent composed of silicon and oxygen. The essential material of the mineral quartz.

silica-alumina ratio The molecules of silicon dioxide (SiO_2) per molecule of aluminum oxide (Al_2O_3) in clay minerals or in soils.

silt 1. A soil separate consisting of particles between 0.05 and 0.002 mm in equivalent diameter. *See* soil separates. 2. A soil textural class.

silting The deposition of water-borne sediments in stream channels, lakes, reservoirs, or on floodplains, usually resulting from a decrease in the velocity of the water.

silt loam A soil textural class containing a large amount of silt and small quantities of sand and clay.

silty clay A soil textural class containing a relatively large amount of silt and clay and a small amount of sand.

silty clay loam A soil textural class containing a relatively large amount of silt, a lesser quantity of clay, and a still smaller quantity of sand.

single-grain structure (obsolete) A soil structure classification in which the soil particles occur almost completely as individual or primary particles with essentially no secondary particles or aggregates being present. (Usually found only in extremely coarse-textured soils.)

site 1. In ecology, an area described or defined by its biotic, climatic, and soil conditions as related to its capacity to produce vegetation. 2. An area sufficiently uniform in biotic, climatic, and soil conditions to produce a particular climax vegetation.

site index 1. A quantitative evaluation of the productivity of a soil for forest growth under the existing or specified environment. 2. The height in feet of the dominant forest vegetation taken at or calculated to an index age, usually 50 or 100 years.

slick spots Small areas in a field that are slick when wet, due to a high content of alkali or of exchangeable sodium.

slope The incline of the surface of a soil. It is usually expressed in percentage of slope, which equals the number of feet of fall per 100 feet of horizontal distance.

soil The collection of natural bodies on the earth's surface, containing living matter, and supporting or capable of supporting plants. Its upper limit is air or water and at its lateral margins it grades to deep water or barren areas of rock or ice. Its lower limit is normally considered to be the lower limit of the common rooting of the native perennial plants, a boundary that is shallow in deserts and tundra and deep in the humid tropics.

soil air The soil atmosphere; the gaseous phase of the soil; it is that volume not occupied by solid or liquid.

soil alkalinity The degree or intensity of alkalinity of a soil, expressed by a value of greater than 7.0 on the pH scale.

soil association 1. A group of defined and named taxonomic soil units occurring together in an individual and characteristic pattern over a geographic region, comparable to plant associations in many ways. (Sometimes called *natural land type.*) 2. A mapping unit used on general soil maps, in which two or more defined taxonomic units occurring together in a characteristic pattern are combined because the scale of the map or the purpose for which it is being made does not require delineation of the individual soils.

soil complex A mapping unit used in detailed soil surveys where two or more defined taxonomic units are so intimately intermixed geographically that it is undesirable or impractical, because of the scale being used, to separate them. A more intimate mixing of smaller areas of individual taxonomic units than that described under soil association.

soil conservation 1. Protection of the soil against physical loss by erosion or against chemical deterioration; that is, excessive loss of fertility by either natural or artificial means. 2. A combination of all management and land use methods which safeguard the soil against depletion or deterioration by natural or by man-induced factors. 3. A division of soil science concerned with soil conservation.

soil extract The solution separated from a soil suspension or from a soil by filtration, centrifugation, suction, or pressure. (May or may not be heated prior to separation.)

soil-formation factors The variable, usually interrelated, natural agencies that are active in and responsible for the formation of soil. The factors are usually grouped into five major categories as follows: parent rock, climate, organisms, topography, and time.

soil genesis 1. The mode of origin of the soil with special reference to the processes or soil-forming factors responsible for the development of the solum or true soil, from the unconsolidated parent material. 2. A division of soil science concerned with soil genesis.

soil horizon A layer of soil or soil material approximately parallel to the land surface and differing from adjacent genetically related layers in physical, chemical, and biological properties, or characteristics such as color, structure, texture, consistency, kinds and numbers of organisms present, degree of acidity or alkalinity, etc. The following table lists the designations and properties of the major soil horizons. Very few if any soils have all of these horizons well developed but every soil has some of them.

Horizon Designation	*Description*
0	Organic horizons of mineral soils.
01	Organic horizons in which essentially the original form of most vegetative matter is visible to the naked eye.
02	Organic horizons in which the original form of most plant or animal matter cannot be recognized with the naked eye.

Horizon Designation	*Description*
A	Mineral horizons consisting of: (1) horizons of organic-matter accumulation formed or forming at or adjacent to the surface; (2) horizons that have lost clay, iron, or aluminum with resultant concentration of quartz or other resistant minerals of sand or silt size; or (3) horizons dominated by (1) or (2) above but transitional to an underlying B or C.
Ap	The plowed portion of the A horizon.
A1	Mineral horizons, formed or forming at or adjacent to the surface, in which the feature emphasized is an accumulation of humified organic matter intimately associated with the mineral fraction.
A2	Mineral horizons in which the feature emphasized is loss of clay, iron, or aluminum, with resultant concentration of quartz or other resistant minerals in sand and silt sizes.
A3	A transitional horizon between A and B, and dominated by properties characteristic of an overlying A1 or A2 but having some subordinate properties of an underlying B.
AB	A horizon transitional between A and B, having an upper part dominated by properties of B, and the two parts cannot be conveniently separated into A3 and B1.
A and B	Horizons that would qualify for A2 except for included parts constituting less than 50 percent of the volume that would qualify as B.
AC	A horizon transitional between A and C, having subordinate properties of both A and C, but not dominated by properties characteristic of either A or C.
B and A	Any horizon qualifying as B in greater than 50 percent of its volume including parts that qualify as A2.

Horizon Designation	*Description*
B	Soil horizon beneath the A horizon. Clay, nutrients, etc., have accumulated in this horizon.
B1	A transitional horizon between B and A1 or between B and A2 in which the horizon is dominated by properties of an underlying B2 but has some subordinate properties of an overlying A1 or A2.
B2	That part of the B horizon where the properties on which the B is based are without clearly expressed subordinate characteristics indicating that the horizon is transitional to an adjacent overlying A or an adjacent underlying C or R.
B3	A transitional horizon between B and C or R in which the properties diagnostic of an overlying B2 are clearly expressed but are associated with clearly expressed properties characteristic of C or R.
C	A mineral horizon or layer, excluding bedrock, that is either like or unlike the material from which the solum is presumed to have formed, relatively little affected by pedogenic processes, and lacking properties diagnostic of A or B.
R	Underlying consolidated bedrock, such as granite, sandstone, or limestone.

soil management 1. The sum total of all tillage operations, cropping practices, fertilizer, lime, and other treatments conducted on or applied to a soil for the production of plants. 2. A division of soil science concerned with the items listed under 1.

soil map A map showing the distribution of soil types or other soil mapping units in relation to the prominent physical and cultural features of the earth's surface.

soil microbiology A subspecialization of soil science concerned with soil-inhabiting microorganisms and with their relation to agriculture, including both plant and animal growth.

soil mineral 1. Any mineral that occurs as a part of soil or in the soil. 2. A natural inorganic compound with definite physical, chemical, and crystalline properties that occurs in the soil.

soil mineralogy A subspecialization of soil science concerned with the homogeneous inorganic materials found in the earth's crust to the depth of weathering or of sedimentation.

soil monolith A vertical section of a soil profile removed from the soil and mounted for display or study.

soil morphology 1. The physical composition, particularly the structural properties, of a soil profile as exhibited by the kinds, thickness, and arrangement of the horizons in the profile, and by the texture, structure, consistency, and porosity of each horizon. 2. The structural characteristics of the soil or any of its parts.

soil organic matter The organic fraction of the soil; includes plant and animal residues at various stages of decomposition, cells and tissues of soil organisms, and substances synthesized by the soil population. Usually determined on soils which have been sieved through a 2.0-mm sieve.

soil pores That part of the bulk volume of soil not occupied by soil particles.

soil science That science dealing with soils as a natural resource on the surface of the earth, including soil formation, classification and mapping, and the physical, chemical, biological, and fertility properties of soils per se; and these properties in relation to their management for crop production.

soil separates Mineral particles, less than 2.0 mm in equivalent diameter, ranging between specified size limits. The names and size limits of separates recognized in the United States are: very coarse sand, 2.0 to 1.0 mm; coarse sand, 1.0 to 0.5 mm; medium sand, 0.5 to 0.25 mm; fine sand, 0.25 to 0.10 mm; very fine sand, 0.10 to 0.05 mm; silt, 0.05 to 0.002 mm; and clay less than 0.002 mm.

soil series The basic unit of soil classification being a subdivision of a family and consisting of soils which are essentially alike in all major profile characteristics.

soil solution The aqueous liquid phase of the soil and its solutes consisting of ions dissociated from the surfaces of the soil particles and of other soluble materials.

soil structure The combination or arrangement of primary soil particles into secondary particles, units, or peds. These secondary units may be, but usually are not, arranged in the profile in such a manner as to give a distinctive characteristic pattern. The secondary units are characterized and classified on the basis of size, shape, and degree of distinctness into classes, types, and grades, respectively.

soil structure grades A grouping or classification of soil structure on the basis of structural strength.

soil survey The systematic examination, description, classification, and mapping of soils in an area. Soil surveys are classified according to the kind and intensity of field examination.

soil texture The relative proportions of the various soil separates in a soil.

soil type (obsolete) The lowest unit in the 1938 system of soil classification; a subdivision of a soil series and consisting of or describing soils that are alike in all characteristics including the texture of the A horizon.

solum (plural: sola) The upper and most weathered part of the soil profile; the A and B horizons.

spoil bank Rock waste, banks, and dumps, from the excavation of ditches.

stones Rock fragments greater than 10 inches in diameter if rounded, and 15 inches along the greater axis if flat.

stratified Arranged in or composed of strata or layers.

strip cropping The practice of growing crops which require different types of tillage, such as row and sod, in alternate strips along contours or across the prevailing direction of wind.

stubble mulch The stubble of crops or crop residues left essentially in place on the land as a surface cover before and during the preparation of the seedbed and at least partly during the growing of a succeeding crop.

subsoil The B horizons of soils with distinct profiles. In soils with weak profile development, the subsoil can be defined as the soil below the plowed soil (or its equivalent) of surface soil), in which roots normally grow. Although a common term, it cannot be defined accurately. It has been carried over from early days when soil was conceived only as the plowed soil and that under it was the subsoil.

subsoiling Breaking of compact subsoils, without inverting them, with a special knifelike instrument (chisel) which is pulled through the soil at depths usually of 12 to 24 inches and at spacings usually of 2 to 5 feet.

substrate 1. That which is laid or spread under; an underlying layer, such as the subsoil. 2. The substance, base, or nutrient on which an organism grows. 3. The compounds or substances that are acted upon by enzymes or catalysts and changed to other compounds in the chemical reaction.

subsurface tillage Tillage with a special sweeplike plow or blade which is drawn beneath the surface at depths of several inches and cuts plant roots and loosens the soil without inverting it or without incorporating the surface cover.

surface soil The uppermost part of the soil, ordinarily moved in tillage, or its equivalent in uncultivated soils and ranging in depth from 3 or 4 inches to 8 or 10. Frequently designated as the plow layer, the Ap layer, or the Ap horizon.

swamp An area saturated with water throughout much of the year but with the surface of the soil usually not deeply submerged. Usually characterized by tree or shrub vegetation.

symbiosis The living together in intimate association of two dissimilar organisms, the cohabitation being mutually beneficial.

tensiometer A device for measuring the negative pressure (or tension) of water in soil in situ; a porus, permeable ceramic cup connected through a tube to a manometer or vacuum gauge.

terrace 1. A level, usually narrow, plain bordering a river, lake, or the sea. Rivers sometimes are bordered by terraces at different levels. 2. A raised, more or less level or horizontal strip of earth usually constructed on or nearly on a contour and supported on the downslope side by rocks or other similar barriers and designed to make the land suitable for tillage and to prevent accelerated erosion. For example, the ancient terraces built by the Incas in the Andes. 3. An embankment with the uphill side sloping toward and into a channel for conducting water, and the downhill side having a relatively sharp decline; constructed across the direction of the slope for the purpose of conducting water from the area above the terrace at a regulated rate of flow and to prevent the accumulation of large volumes of water on the downslope side of cultivated fields. The depth of the channel, the width of the terrace ridge, and the spacings of the terraces on a field are varied with soil types, cropping systems, climatic conditions, and other factors.

tidal flats Areas of nearly flat, barren mud periodically covered by tidal waters. Normally these materials have an excess of soluble salt.

tight soil A compact, relatively impervious and tenacious soil (or subsoil) which may or may not be plastic.

tile drain Concrete, plastic, or ceramic pipe placed at suitable depths and spacings in the soil or subsoil to provide water outlets from the soil.

till 1. Unstratified glacial drift deposited directly by the ice and consisting of clay, sand, gravel, and boulders intermingled in any proportion. 2. To plow and prepare for seeding; to seed or cultivate the soil.

tilth The physical condition of soil as related to its ease of tillage, fitness as a seedbed, and its impedance to seedling emergence and root penetration.

toposequence A sequence of related soils that differ, one from the other, primarily because of topography as a soil-formation factor.

topsoil 1. The layer of soil moved in cultivation. *See* surface soil. 2. The A horizon. 3. The A1 horizon. 4. Presumably fertile soil material used to top-dress roadbanks, gardens, and lawns.

trace elements (obsolete) *See* micronutrient.

transpiration Loss of water vapor from the leaves and stems of living plants to the atmosphere.

truncated Having lost all or part of the upper soil horizon or horizons.

tundra A level or undulating, treeless plain characteristic of arctic regions.

Ultisols Old, well-developed soils showing the ultimate degree of soil formation.

unsaturated flow The movement of water in a soil which is not filled to capacity with water.

upland soils High ground; ground elevated above the lowlands along the rivers or between hills.

urban land Areas so altered or obstructed by urban works or structures that identification of soils is not feasible. A miscellaneous land type.

valence The combining capacity or electrical charge of atoms or groups of atoms. Sodium (Na^+) and potassium (K^+) are monovalent, while calcium (Ca^{++}) is divalent.

value, color The relative lightness or intensity of color and approximately a function of the square root of the total amount of light. One of the three variables of color.

virgin soil A soil that has not been significantly disturbed from its natural environment.

viscosity, of fluid Property of stickiness of liquid or gas due to its cohesive and adhesive characteristics.

volatilization The evaporation or changing of a substance from liquid to vapor.

volume weight (obsolete) *See* bulk density.

wasteland Land not suitable for, or capable of, producing materials or services of value. A miscellaneous land type.

waterlogged Saturated with water.

water table The upper surface of ground water or that level below which the soil is saturated with water; locus of points in soil water at which the hydraulic pressure is equal to atmospheric pressure.

water table, perched The water table of a saturated layer of soil which is separated from an underlying saturated layer by an unsaturated layer.

watershed In the U.S. the term refers to the total area above a given point on a stream that contributes water to the flow at that point. Synonyms are *drainage basin* or *catchment basin.* In some other countries, the term is used for the topographic boundary separating one drainage basin from another.

weathering All physical and chemical changes produced in rocks, at or near the earth's surface, by atmospheric agents.

wilting point *Also called* permanent wilting percentage. That soil moisture level at which plants wilt and cannot be revived by placing them in a saturated atmosphere, i.e., soil moisture levels at which plants wilt and die.

windbreak A planting of trees, shrubs, or other vegetation, usually perpendicular or nearly so to the principal wind direction, to protect soil, crops, homesteads, roads, etc., against the effects of winds, such as wind erosion and the drifting of soil and snow.

xerophytes Plants that grow in or on extremely dry soils or soil materials.

yield, sustained A continual annual, or periodic, yield of plants or plant material from an area; implies management practices which will maintain the productive capacity of the land.

zymogenous flora Organisms found in soils in large numbers immediately following the addition of readily decomposable organic materials.

Index